Thomas Henry Huxley

On our knowledge of the causes of the phenomena of organic nature. Being six lectures to working men, delivered at the Museum of practical geology

Thomas Henry Huxley

On our knowledge of the causes of the phenomena of organic nature. Being six lectures to working men, delivered at the Museum of practical geology

ISBN/EAN: 9783337220457

Printed in Europe, USA, Canada, Australia, Japan

Cover: Foto ©berggeist007 / pixelio.de

More available books at **www.hansebooks.com**

ON OUR KNOWLEDGE

OF THE

CAUSES OF THE PHENOMENA

OF

ORGANIC NATURE.

BEING

SIX LECTURES TO WORKING MEN,

DELIVERED AT THE

MUSEUM OF PRACTICAL GEOLOGY.

BY

PROFESSOR HUXLEY, F.R.S.

LONDON:
ROBERT HARDWICKE, 192, PICCADILLY.
1863.

NOTICE.

The Publisher of these interesting Lectures, having made an arrangement for their publication with Mr. J. A. Mays, the Reporter, begs to append the following note from Professor Huxley:—

"Mr. J. Aldous Mays, who is taking shorthand notes of my 'Lectures to Working Men,' has asked me to allow him, on his own account, to print those Notes for the use of my audience. I willingly accede to this request, on the understanding that a notice is prefixed to the effect that I have no leisure to revise the Lectures, or to make alterations in them, beyond the correction of any important error in a matter of fact.

"T. H. HUXLEY."

CONTENTS.

NO.		PAGE
I.	The Present Condition of Organic Nature	5
II.	The Past Condition of Organic Nature	28
III.	The Method by which the Causes of the Present and Past Conditions of Organic Nature are to be Discovered.— The Origination of Living Beings	52
IV.	The Perpetuation of Living Beings, Hereditary Transmission and Variation	82
V.	The Conditions of Existence as affecting the Perpetuation of Living Beings	106
VI.	A Critical Examination of the Position of Mr. Darwin's Work, "On the Origin of Species," in relation to the complete Theory of the Causes of the Phenomena of Organic Nature	132

LECTURE I.

THE PRESENT CONDITION OF ORGANIC NATURE.

WHEN it was my duty to consider what subject I would select for the six lectures which I shall now have the pleasure of delivering to you, it occurred to me that I could not do better than endeavour to put before you in a true light, or in what I might perhaps with more modesty call, that which I conceive myself to be the true light, the position of a book which has been more praised and more abused, perhaps, than any book which has appeared for some years;—I mean Mr. Darwin's work on the "Origin of Species." That work, I doubt not, many of you have read; for I know the inquiring spirit which is rife among you. At any rate, all of you will have heard of it,—some by one kind of report and some by another kind of report; the attention of all and the curiosity of all have been probably more or less excited on the subject of that work. All I can do, and all I shall attempt to do, is to put before you that kind of judgment which has been formed by a man, who, of course, is liable to judge erroneously; but at any rate, of one whose business and profession it is to form judgments upon questions of this nature.

And here, as it will always happen when dealing with an extensive subject, the greater part of my course—if, indeed, so small a number of lectures can be properly called a course — must be devoted to preliminary matters, or rather to a statement of those facts and of those principles which the work itself dwells upon, and brings more or less directly before us. I have no right to suppose that all or any of you are naturalists; and even if you were, the misconceptions and misunderstandings prevalent even among naturalists on these matters would make it desirable that I should take the course I now propose to take,—that I should start from the beginning,—that I should endeavour to point out what is the existing state of the organic world—that I should point out its past condition,—that I should state what is the precise nature of the undertaking which Mr. Darwin has taken in hand; that I should endeavour to show you what are the only methods by which that undertaking can be brought to an issue, and to point out to you how far the author of the work in question has satisfied those conditions, how far he has not satisfied them, how far they are satisfiable by man, and how far they are not satisfiable by man.

To-night, in taking up the first part of the question, I shall endeavour to put before you a sort of broad notion of our knowledge of the condition of the living world. There are many ways of doing this. I might deal with it pictorially and graphically. Following the example of Humboldt in his "Aspects of Nature," I might endeavour to point out the infinite variety of organic life in every mode of its existence, with reference to the variations of climate and the like; and such an attempt would be fraught with interest to us all; but considering the subject before us, such a course would not be that best calculated to assist us. In an argument of this kind we must go further and dig deeper into the matter; we must endeavour to look into the foundations of living Nature, if I may so say, and discover the principles involved in some of her most secret operations. I propose, therefore, in the first place, to take some ordinary animal with which you are all familiar, and, by easily comprehensible and obvious examples drawn from it, to show what are the kind of problems which living beings in general lay before us; and I shall then show you that the same problems are laid open to us by all kinds of living beings. But, first, let me say in what sense I have used the words "organic nature." In speaking of the causes which lead to our present knowledge of organic nature, I have used it almost as an equivalent of the word "living," and for this reason,—that in almost all living beings you can distinguish several distinct portions set apart to do particular things and work in a particular way. These are termed "organs," and the whole together is called "organic." And as it is universally characteristic of them, the term "organic" has been very conveniently employed to denote the whole of living nature,—the whole of the plant world, and the whole of the animal world.

Few animals can be more familiar to you than that whose skeleton is shown on our diagram. You need not bother yourselves with this "*Equus caballus*" written under it; that is only the Latin name of it, and does not make it any better. It simply means the common Horse. Suppose we wish to understand all about the Horse. Our first object must be to study the structure of the animal. The whole of his body is inclosed within a hide, a skin covered with hair; and if that hide or skin be taken off, we find a great mass of flesh, or what is technically called muscle, being the substance which by its power of contraction enables the animal to move. These muscles move the hard parts one upon the other, and so give that strength and power of motion which renders the Horse so useful to us in the performance of those services in which we employ him.

And then, on separating and removing the whole of this skin and flesh, you have a great series of bones, hard structures, bound together with ligaments, and forming the skeleton which is represented here.

In that skeleton there are a number of parts to be recognized. The long series of bones, beginning from the skull and ending in the tail, is called the spine, and those in front are the ribs; and then there are two pairs of limbs, one before and one behind; and there are what we all know as the fore-legs and the hind-legs. If we pursue our researches into the interior of this animal, we find within the framework of the skeleton a great cavity, or rather, I should say, two great cavities,—one cavity beginning in the skull and running through the neck-bones, along the spine, and ending in the tail, containing the brain and the spinal marrow, which are extremely important organs. The second great cavity, commencing with the mouth, contains the gullet, the stomach, the long intestine, and all the rest of those internal apparatus which are essential for digestion;

and then in the same great cavity, there are lodged the heart and all the great vessels going from it; and, besides that, the organs of respiration—the lungs; and then the kidneys, and the organs of reproduction, and so on. Let us now endeavour to reduce this notion of a horse that we now have, to some such kind of simple expression as can be at once, and without difficulty, retained in the mind, apart from all minor details. If I make a transverse section, that is, if I were to saw a dead horse across, I should find that, if I left out the details, and supposing I took my section through the anterior region, and through the fore-limbs, I should have here this kind of section of the body (Fig. 1). Here would be the upper part of the animal—that great mass of bones that we spoke of as the spine (*a*, Fig. 1.) Here I should have the alimentary canal (*b*, Fig. 1). Here I should have the heart (*c*, Fig. 1); and then you see, there would be a kind of

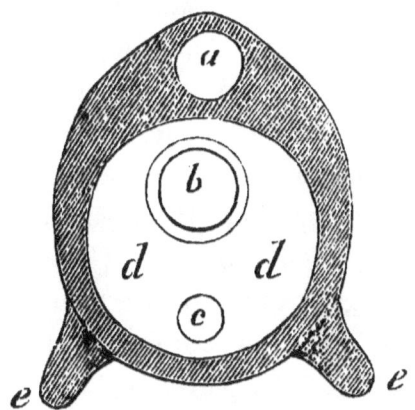

Fig. 1.

double tube, the whole being inclosed within the hide; the spinal marrow would be placed in the upper tube (*a*, Fig. 1), and in the lower tube (*d d*, Fig. 1), there would be the alimentary canal (*b*), and the heart (*c*); and here I shall have the legs proceeding from each side. For simplicity's sake, I represent them merely as stumps (*e e*, Fig.1). Now that is a horse—as mathematicians would say — reduced to its most simple expression. Carry that in your minds, if you please, as a simplified idea of the structure of the Horse. The considerations which I have now put before you belong to what we technically call the 'Anatomy' of the Horse. Now, suppose we go to work upon these several parts,—flesh and hair, and skin and bone, and lay open these various organs with our scalpels, and examine them by means of our magnifying-glasses, and see what we can make of them. We shall find that the flesh is made up of bundles of strong fibres. The brain and nerves, too, we shall find, are made up of fibres, and these queer-looking things that are called ganglionic corpuscles. If we take a slice of the bone and examine it, we shall find that it is very like this diagram of a section of the bone of an ostrich, though differing, of course, in some details; and if we take any part whatsoever of the tissue, and examine it, we shall find it all has a minute structure, visible only under the microscope. All these parts constitute microscopic anatomy or 'Histology.' These parts are constantly being changed; every part is constantly growing, decaying, and being replaced during the life of the animal. The tissue is constantly replaced by new material; and if you go back to the young state of the tissue in the case of muscle, or in the case of skin, or any of the organs I have mentioned, you will find that they all come under the same condition. Every one of these microscopic filaments and fibres (I now speak merely of the general character of the whole process)—every one of these parts—could be traced down to some modification of a tissue which can be readily divided into little particles of fleshy matter, of that substance which is composed of the chemical elements, carbon, hydrogen, oxygen, and nitrogen, having such a shape as this (Fig. 2). These particles, into which all primitive tissues break up, are called cells. If I were to make a section of a piece of the skin of my hand, I should find that it was made up of these cells. If I examine the fibres which form the various organs of all living animals, I should find that all of them, at one time or other, had been formed out of a substance consisting of similar elements; so that you see, just as we reduced the whole body in the gross to that sort of simple expression given in Fig. 1, so we may reduce the whole of the microscopic structural elements to a form of even greater simplicity; just as the plan of the whole body may be so represented in a sense (Fig. 1), so the primary structure of every tissue may be represented by a mass of cells (Fig. 2).

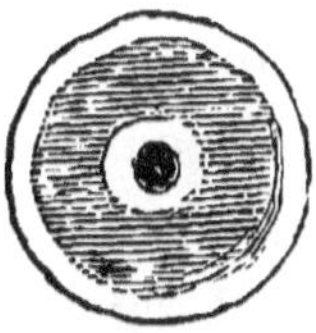

FIG. 2.

Having thus, in this sort of general way, sketched to you what I may call, perhaps, the architecture of the body of the Horse, (what we term technically its Morphology,) I must now turn to another aspect. A horse is not a mere dead structure: it is an active, living, working machine. Hitherto we have, as it were, been looking at a steam-engine with the fires out, and nothing in the boiler; but the body of the living animal is a beautifully-formed active machine, and every part has its different work to do in the working of that machine, which is what we call its life. The Horse, if you see him after his day's work is done, is cropping the grass in the fields, as it may be, or munching the oats in his stable. What is he doing? His jaws are working as a mill—and a very complex mill too—grinding the corn, or crushing the grass to a pulp. As soon as that operation has taken place, the food is passed down to the stomach, and there it is mixed with the chemical fluid called the gastric juice, a substance which has the peculiar property of making soluble and dissolving out the nutritious matter in the grass, and leaving behind those parts which are not nutritious; so that you have, first, the mill, then a sort of chemical digester; and then the food, thus partially dissolved, is carried back by the muscular contractions of the intestines into the hinder parts of the body, while the soluble portions are taken up into the blood. The blood is contained in a vast system of pipes, spreading through the whole body, connected with a force-pump,—the heart,—which, by its position and by the contractions of its valves, keeps the blood constantly circulating in one direction, never allowing it to rest; and then, by means of this circulation of the blood, laden as it is with the products of digestion, the skin, the flesh, the hair, and every other part of the body, draws from it that which it wants, and every one of these organs derives those materials which are necessary to enable it to do its work.

The action of each of these organs, the performance of each of these various duties, involve in their operation a continual absorption of the matters necessary for their support, from the blood, and a constant formation of waste products, which are returned to the blood, and conveyed by it to the lungs and the kidneys, which are organs that have allotted to them the office of extracting, separating, and getting rid of these waste products;

and thus the general nourishment, labour, and repair of the whole machine is kept up with order and regularity. But not only is it a machine which feeds and appropriates to its own support the nourishment necessary to its existence—it is an engine for locomotive purposes. The Horse desires to go from one place to another; and to enable it to do this, it has those strong contractile bundles of muscles attached to the bones of its limbs, which are put in motion by means of a sort of telegraphic apparatus formed by the brain and the great spinal cord running through the spine or backbone; and to this spinal cord are attached a number of fibres termed nerves, which proceed to all parts of the structure. By means of these the eyes, nose, tongue, and skin—all the organs of perception—transmit impressions or sensations to the brain, which acts as a sort of great central telegraph-office, receiving impressions and sending messages to all parts of the body, and putting in motion the muscles necessary to accomplish any movement that may be desired. So that you have here an extremely complex and beautifully-proportioned machine, with all its parts working harmoniously together towards one common object—the preservation of the life of the animal.

Now, note this: the Horse makes up its waste by feeding, and its food is grass or oats, or perhaps other vegetable products; therefore, in the long run, the source of all this complex machinery lies in the vegetable kingdom. But where does the grass, or the oat, or any other plant, obtain this nourishing food-producing material? At first it is a little seed, which soon begins to draw into itself from the earth and the surrounding air matters which in themselves contain no vital properties whatever; it absorbs into its own substance water, an inorganic body; it draws into its substance carbonic acid, an inorganic matter; and ammonia, another inorganic matter, found in the air; and then, by some wonderful chemical process, the details of which chemists do not yet understand, though they are near foreshadowing them, it combines them into one substance, which is known to us as 'Protein,' a complex compound of carbon, hydrogen, oxygen, and nitrogen, which alone possesses the property of manifesting vitality and of permanently supporting animal life. So that, you see, the waste products of the animal economy, the effete materials which are continually being thrown off by all living beings, in the form of organic matters, are constantly replaced by supplies of the necessary repairing and rebuilding materials drawn from the plants, which in their turn manufacture them, so to speak, by a mysterious combination of those same inorganic materials.

Let us trace out the history of the Horse in another direction. After a certain time, as the result of sickness or disease, the effect of accident, or the consequence of old age, sooner or later, the animal dies. The multitudinous operations of this beautiful mechanism flag in their performance, the Horse loses its vigour, and after passing through the curious series of changes comprised in its formation and preservation, it finally decays, and ends its life by going back into that inorganic world from which all but an inappreciable fraction of its substance was derived. Its bones become mere carbonate and phosphate of lime; the matter of its flesh, and of its other parts, becomes, in the long run, converted into carbonic acid, into water, and into ammonia. You will now, perhaps, understand the curious relation of the animal with the plant, of the organic with the inorganic world, which is shown in this diagram.

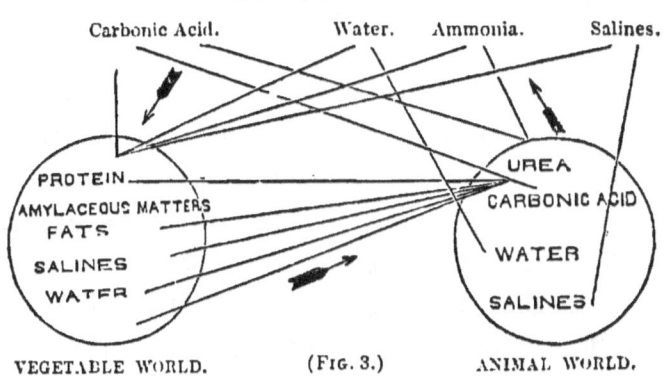

(Fig. 3.)

The plant gathers these inorganic materials together and makes them up into its own substance. The animal eats the plant and appropriates the nutritious portions to its own sustenance, rejects and gets rid of the useless matters; and, finally, the animal itself dies, and its whole body is decomposed and returned into the inorganic world. There is thus a constant circulation from one to the other, a continual formation of organic life from inorganic matters, and as constant a return of the matter of living bodies to the inorganic world; so that the materials of which our bodies are composed are largely, in all probability, the substances which constituted the matter of long extinct creations, but which have in the interval constituted a part of the inorganic world.

Thus we come to the conclusion, strange at first sight, that the MATTER constituting the living world is identical with that which forms the inorganic world. And not less true is it that, remarkable as are the powers or, in other words, as are the FORCES which are exerted by living beings, yet all these forces are either identical with those which exist in the inorganic world, or they are convertible into them; I mean in just the same sense as the researches of physical philosophers have shown that heat is convertible into electricity, that electricity is convertible into magnetism, magnetism into mechanical force or chemical force, and any one of them with the other, each being measurable in terms of the other,—even so, I say, that great law is applicable to the living world. Consider why is the skeleton of this horse capable of supporting the masses of flesh and the various organs forming the living body, unless it is because of the action of the same forces of cohesion which combines together the particles of matter composing this piece of chalk? What is there in the muscular contractile power of the animal but the force which is expressible, and which is in a certain sense convertible, into the force of gravity which it overcomes? Or, if you go to more hidden processes, in what does the process of digestion differ from those processes which are carried on in the laboratory of the chemist? Even if we take the most recondite and most complex operations of animal life—those of the nervous system, these of late years have been shown to be—I do not say identical in any sense with the electrical processes—but this has been shown, that they are in some way or other associated with them; that is to say, that every amount of nervous action is accompanied by a certain amount of electrical disturbance in the particles of the nerves in which that nervous action is carried on. In this way the nervous action is related to electricity in the same way that heat is related to electricity; and the same sort of argument which demonstrates the two latter to be related to one another shows that the nervous forces are correlated to electricity; for the experiments of M. Dubois Reymond and others have shown that whenever a nerve is in a state of excitement, sending a message to the muscles or conveying an impression to the brain, there is a disturbance of the electrical condition of that nerve which does not exist at other times; and there are a number of other facts and phenomena of that sort; so that we come to the broad conclusion that not only as to living matter itself, but as to the forces that matter exerts, there is a close relationship between the organic and the inorganic world—the difference between them arising from the diverse combination and disposition of identical forces, and not from any primary diversity, so far as we can see.

I said just now that the Horse eventually died and became converted into the same inorganic substances from whence all but an inappreciable fraction of its substance demonstrably originated, so that the actual wanderings of matter are as remarkable as the transmigrations of the soul fabled by Indian tradition. But before death has occurred, in the one sex or the other, and in fact in both, certain products or parts of the organism have been set free, certain parts of the organisms of the two sexes have come into contact with one another, and from that conjunction, from that union which then takes place, there results the formation of a new being. At stated times the mare, from a particular part of the interior of her body, called the ovary, gets rid of a minute particle of matter comparable in all essential respects with that which we called a cell a little while since, which cell contains a kind of nucleus in its centre, surrounded by a clear space and by a viscid mass of protein substance (Fig. 2); and though it is different in appearance from the eggs which we are mostly acquainted with, it is really an egg. After a time this minute particle of matter, which may only be a small fraction of a grain in weight, undergoes a series of changes,—wonderful, complex changes. Finally, upon its surface there is fashioned a little elevation, which afterwards becomes divided and marked by a groove. The lateral boundaries of the groove extend upwards and downwards, and at length give rise to a double tube. In the upper and smaller tube the spinal marrow and brain are fashioned; in the lower, the alimentary canal and heart; and at length two pairs of buds shoot out at the sides of the body, and they are the rudiments of the limbs. In fact a true drawing of a section of the embryo in this state would in all essential respects resemble that diagram of a horse reduced to its simplest expression, which I first placed before you (Fig. 1).

Slowly and gradually these changes take place. The whole of the body, at first, can be broken up into "cells," which become in one place metamorphosed into muscle, —in another place into gristle and bone,—in another place into fibrous tissue,—and in another into hair;

every part becoming gradually and slowly fashioned, as if there were an artificer at work in each of these complex structures that I have mentioned. This embryo, as it is called, then passes into other conditions. I should tell you that there is a time when the embryos of neither dog, nor horse, nor porpoise, nor monkey, nor man, can be distinguished by any essential feature one from the other; there is a time when they each and all of them resemble this one of the Dog. But as development advances, all the parts acquire their speciality, till at length you have the embryo converted into the form of the parent from which it started. So that you see, this living animal, this horse, begins its existence as a minute particle of nitrogenous matter, which, being supplied with nutriment (derived, as I have shown, from the inorganic world), grows up according to the special type and construction of its parents, works and undergoes a constant waste, and that waste is made good by nutriment derived from the inorganic world; the waste given off in this way being directly added to the inorganic world. Eventually the animal itself dies, and, by the process of decomposition, its whole body is returned to those conditions of inorganic matter in which its substance originated.

This, then, is that which is true of every living form, from the lowest plant to the highest animal—to man himself. You might define the life of every one in exactly the same terms as those which I have now used; the difference between the highest and the lowest being simply in the complexity of the developmental changes, the variety of the structural forms, and the diversity of the physiological functions which are exerted by each.

If I were to take an oak tree, as a specimen of the plant world, I should find that it originated in an acorn, which, too, commenced in a cell; the acorn is placed in the ground, and it very speedily begins to absorb the inorganic matters I have named, adds enormously to its bulk, and we can see it, year after year, extending itself upward and downward, attracting and appropriating to itself inorganic materials, which it vivifies, and eventually, as it ripens, gives off its own proper acorns, which again run the same course. But I need not multiply examples,—from the highest to the lowest the essential features of life are the same, as I have described in each of these cases.

So much, then, for these particular features of the organic world, which you can understand and comprehend, so long as you confine yourself to one sort of living being, and study that only.

But, as you know, horses are not the only living creatures in the world; and again, horses, like all other animals, have certain limits—are confined to a certain area on the surface of the earth on which we live,—and, as that is the simpler matter, I may take that first. In its wild state, and before the discovery of America, when the natural state of things was interfered with by the Spaniards, the Horse was only to be found in parts of the earth which are known to geographers as the Old World; that is to say, you might meet with horses in Europe, Asia, or Africa; but there were none in Australia, and there were none whatsoever in the whole continent of America, from

Labrador down to Cape Horn. This is an empirical fact, and it is what is called, stated in the way I have given it you, the 'Geographical Distribution' of the Horse.

Why horses should be found in Europe, Asia, and Africa, and not in America, is not obvious; the explanation that the conditions of life in America are unfavourable to their existence, and that, therefore, they had not been created there, evidently does not apply; for when the invading Spaniards, or our own yeomen farmers, conveyed horses to these countries for their own use, they were found to thrive well and multiply very rapidly; and many are even now running wild in those countries, and in a perfectly natural condition. Now, suppose we were to do for every animal what we have here done for the Horse,—that is, to mark off and distinguish the particular district or region to which each belonged; and supposing we tabulated all these results, that would be called the Geographical Distribution of animals, while a corresponding study of plants would yield as a result the Geographical Distribution of plants.

I pass on from that now, as I merely wished to explain to you what I meant by the use of the term 'Geographical Distribution.' As I said, there is another aspect, and a much more important one, and that is, the relations of the various animals to one another. The Horse is a very well-defined matter-of-fact sort of animal, and we are all pretty familiar with its structure. I dare say it may have struck you, that it resembles very much no other member of the animal kingdom, except perhaps the Zebra or the Ass.

But let me ask you to look along these diagrams. Here is the skeleton of the Horse, and here the skeleton of the Dog. You will notice that we have in the Horse a skull, a backbone and ribs, shoulder-blades and haunch-bones. In the fore-limb, one upper arm-bone, two fore arm-bones, wrist-bones (wrongly called knee), and middle hand-bones, ending in the three bones of a finger, the last of which is sheathed in the horny hoof of the fore-foot: in the hind-limb, one thigh-bone, two leg-bones, ankle-bones, and middle foot-bones, ending in the three bones of a toe, the last of which is encased in the hoof of the hind-foot. Now turn to the Dog's skeleton. We find identically the same bones, but more of them, there being more toes in each foot, and hence more toe-bones.

Well, that is a very curious thing! The fact is that the Dog and the Horse—when one gets a look at them without the outward impediments of the skin—are found to be made in very much the same sort of fashion. And if I were to make a transverse section of the Dog, I should find the same organs that I have already shown you as forming parts of the Horse. Well, here is another skeleton—that of a kind of Lemur—you see he has just the same bones; and if I were to make a transverse section of it, it would be just the same again. In your mind's eye turn him round, so as to put his backbone in a position inclined obliquely upwards and forwards, just as in the next three diagrams, which represent the skeletons of an Orang, a Chimpanzee, and a Gorilla, and you find you have no trouble in identifying the bones throughout; and lastly turn to the end of the series, the diagram representing a man's skeleton, and still you find no great structural feature essentially altered. There are the same bones in the same relations. From the Horse we pass on and on, with gradual steps, until we arrive at last at the highest known forms. On the other hand, take the other line of diagrams, and pass from the Horse downwards in the scale to this fish; and still, though the modifications are vastly greater, the essential framework of the organization remains unchanged. Here, for instance, is a Porpoise; here is its strong backbone, with the cavity running through it, which contains the spinal cord; here are the ribs, here the shoulder-blade; here is the little short upper-arm bone, here are the two forearm bones, the wrist-bone, and the finger-bones.

Strange, is it not, that the Porpoise should have in this queer-looking affair—its flapper (as it is called), the same fundamental elements as the fore-leg of the Horse or the Dog, or the Ape or Man; and here you will notice a very curious thing,—the hinder limbs are absent. Now, let us make another jump. Let us go to the Codfish: here you see is the forearm, in this large pectoral fin—carrying your mind's eye onward from the flapper of the Porpoise. And here you have the hinder limbs restored in the shape of these ventral fins. If I were to make a transverse section of this, I should find just the same organs that we have before noticed. So that, you see, there comes out this strange conclusion as the result of our investigations, that the Horse, when examined and compared with other animals, is found by no means to stand alone in nature; but that there are an enormous number of other creatures which have backbones, ribs, and legs, and other parts arranged in the same general manner, and in all their formation exhibiting the same broad peculiarities.

I am sure that you cannot have followed me even in this extremely elementary exposition of the structural relations of animals, without seeing what I have been driving at all through, which is, to show you that, step by step, naturalists have come to the idea of a unity of plan, or conformity of construction, among animals which appeared at first sight to be extremely dissimilar.

And here you have evidence of such a unity of plan among all the animals which have backbones, and which we technically call *Vertebrata*. But there are multitudes of other animals, such as crabs, lobsters, spiders, and so on, which we term *Annulosa*. In these I could not point out to you the parts that correspond with those of the Horse,—the backbone, for instance,—as they are constructed upon a very different principle, which is also common to all of them; that is to say, the Lobster, the Spider, and the Centipede, have a common plan running through their whole arrangement, in just the same way that the Horse, the Dog, and the Porpoise assimilate to each other.

Yet other creatures—whelks, cuttlefishes, oysters, snails, and all their tribe (*Mollusca*)—resemble one another in the same way, but differ from both *Vertebrata* and *Annulosa*; and the like is true of the animals called *Cœlenterata* (Polypes) and *Protozoa* (animalcules and sponges).

Now, by pursuing this sort of comparison, naturalists

have arrived at the conviction that there are,—some think five, and some seven,—but certainly not more than the latter number—and perhaps it is simpler to assume five — distinct plans or constructions in the whole of the animal world; and that the hundreds of thousands of species of creatures on the surface of the earth, are all reducible to those five, or, at most, seven, plans of organization.

But can we go no further than that? When one has got so far, one is tempted to go on a step and inquire whether we cannot go back yet further and bring down the whole to modifications of one primordial unit. The anatomist cannot do this; but if he call to his aid the study of development, he can do it. For we shall find that, distinct as those plans are, whether it be a porpoise or man, or lobster, or any of those other kinds I have mentioned, every one begins its existence with one and the same primitive form,—that of the egg, consisting, as we have seen, of a nitrogenous substance, having a small particle or nucleus in the centre of it. Furthermore, the earlier changes of each are substantially the same. And it is in this that lies that true "unity of organization" of the animal kingdom which has been guessed at and fancied for many years; but which it has been left to the present time to be demonstrated by the careful study of development. But is it possible to go another step further still, and to show that in the same way the whole of the organic world is reducible to one primitive condition of form? Is there among the plants the same primitive form of organization, and is that identical with that of the animal kingdom? The reply to that question, too, is not uncertain or doubtful. It is now proved that every plant begins its existence under the same form; that is to say, in that of a cell—a particle of nitrogenous matter having substantially the same conditions. So that if you trace back the oak to its first germ, or a man, or a horse, or lobster, or oyster, or any other animal you choose to name, you shall find each and all of these commencing their existence in forms essentially similar to each other: and, furthermore, that the first processes of growth, and many of the subsequent modifications, are essentially the same in principle in almost all.

In conclusion, let me, in a few words, recapitulate the positions which I have laid down. And you must understand that I have not been talking mere theory; I have been speaking of matters which are as plainly demonstrable as the commonest propositions of Euclid— of facts that must form the basis of all speculations and beliefs in Biological science. We have gradually traced down all organic forms, or, in other words, we have analyzed the present condition of animated nature, until we found that each species took its origin in a form similar to that under which all the others commenced their existence. We have found the whole of the vast array of living forms with which we are surrounded, constantly growing, increasing, decaying, and disappearing; the animal constantly attracting, modifying, and applying to its sustenance the matter of the vegetable kingdom, which derived its support from the absorption and conversion of inorganic matter. And so constant and universal is this absorption, waste, and reproduction, that it may be said with perfect certainty that there is left in no one of our bodies at the present moment a millionth part of the matter of which they were originally formed! We have seen, again, that not only is the living matter derived from the inorganic world, but that the forces of that matter are all of them correlative with and convertible into those of inorganic nature.

This, for our present purposes, is the best view of the present condition of organic nature which I can lay before you: it gives you the great outlines of a vast picture, which you must fill up by your own study.

In the next lecture I shall endeavour in the same way to go back into the past, and to sketch in the same broad manner the history of life in epochs preceding our own.

LECTURE II.

THE PAST CONDITION OF ORGANIC NATURE.

In the lecture which I delivered last Monday evening, I endeavoured to sketch in a very brief manner, but as well as the time at my disposal would permit, the present condition of organic nature, meaning by that large title simply an indication of the great, broad, and general principles which are to be discovered by those who look attentively at the phenomena of organic nature as at present displayed. The general result of our investigations might be summed up thus: we found that the multiplicity of the forms of animal life, great as that may be, may be reduced to a comparatively few primitive plans or types of construction; that a further study of the development of those different forms revealed to us that they were again reducible, until we at last brought the infinite diversity of animal, and even vegetable life, down to the primordial form of a single cell.

We found that our analysis of the organic world, whether animals or plants, showed, in the long run, that they might both be reduced into, and were, in fact, composed of the same constituents. And we saw that the plant obtained the materials constituting its substance by a peculiar combination of matters belonging entirely to the inorganic world; that, then, the animal was constantly appropriating the nitrogenous matters of the plant to its own nourishment, and returning them back to the inorganic world, in what we spoke of as its waste; and that, finally, when the animal ceased to exist, the constituents of its body were dissolved and transmitted to that inorganic world whence they had been at first abstracted. Thus we saw in both the blade of grass and the horse but the same elements differently combined and arranged. We discovered a continual circulation going on,—the plant drawing in the elements of inorganic nature and combining them into food for the animal creation; the animal borrowing from the plant the matter for its own support, giving off during its life products which returned immediately to the inorganic world; and that, eventually, the constituent materials of the whole structure of both animals and plants were thus returned to their original source: there was a constant passage from one state of existence to another, and a returning back again.

Lastly, when we endeavoured to form some notion of the nature of the forces exercised by living beings, we discovered that they—if not capable of being subjected to the same minute analysis as the constituents of those beings themselves—that they were correlative with— that they were the equivalents of the forces of inorganic nature—that they were, in the sense in which the term is now used, convertible with them. That was our general result.

And now, leaving the Present, I must endeavour

in the same manner to put before you the facts that are to be discovered in the Past history of the living world, in the past conditions of organic nature. We have, to-night, to deal with the facts of that history —a history involving periods of time before which our mere human records sink into utter insignificance—a history the variety and physical magnitude of whose events cannot even be foreshadowed by the history of human life and human phenomena—a history of the most varied and complex character.

We must deal with the history, then, in the first place, as we should deal with all other histories. The historical student knows that his first business should be to inquire into the validity of his evidence, and the nature of the record in which the evidence is contained, that he may be able to form a proper estimate of the correctness of the conclusions which have been drawn from that evidence. So, here, we must pass, in the first place, to the consideration of a matter which may seem foreign to the question under discussion. We must dwell upon the nature of the records, and the credibility of the evidence they contain; we must look to the completeness or incompleteness of those records themselves, before we turn to that which they contain and reveal. The question of the credibility of the history, happily for us, will not require much consideration, for, in this history, unlike those of human origin, there can be no cavilling, no differences as to the reality and truth of the facts of which it is made up; the facts state themselves, and are laid out clearly before us.

But, although one of the greatest difficulties of

the historical student is cleared out of our path, there are other difficulties—difficulties in rightly interpreting the facts as they are presented to us—which may be compared with the greatest difficulties of any other kinds of historical study.

What is this record of the past history of the globe, and what are the questions which are involved in an inquiry into its completeness or incompleteness? That record is composed of mud; and the question which we have to investigate this evening resolves itself into a question of the formation of mud. You may think, perhaps, that this is a vast step—of almost from the sublime to the ridiculous—from the contemplation of the history of the past ages of the world's existence to the consideration of the history of the formation of mud! But, in nature, there is nothing mean and unworthy of attention; there is nothing ridiculous or contemptible in any of her works; and this inquiry, you will soon see, I hope, takes us to the very root and foundations of our subject.

How, then, is mud formed? Always, with some trifling exception, which I need not consider now—always, as the result of the action of water, wearing down and disintegrating the surface of the earth and rocks with which it comes in contact—pounding and grinding it down, and carrying the particles away to places where they cease to be disturbed by this mechanical action, and where they can subside and rest. For the ocean, urged by winds, washes, as we know, a long extent of coast, and every wave, loaded as it is with particles of sand and gravel as it breaks upon the shore, does something towards the disintegrating process. And thus, slowly but surely, the hardest rocks are gradually ground down to a powdery substance; and the mud thus formed, coarser or finer, as the case may be, is carried by the rush of the tides, or currents, till it reaches the comparatively deeper parts of the ocean, in which it can sink to the bottom, that is, to parts where there is a depth of about fourteen or fifteen fathoms, a depth at which the water is, usually, nearly motionless, and in which, of course, the finer particles of this detritus, or mud as we call it, sinks to the bottom.

Or, again, if you take a river, rushing down from its mountain sources, brawling over the stones and rocks that intersect its path, loosening, removing, and carrying with it in its downward course the pebbles and lighter matters from its banks, it crushes and pounds down the rocks and earths in precisely the same way as the wearing action of the sea waves. The matters forming the deposit are torn from the mountain-side and whirled impetuously into the valley, more slowly over the plain, thence into the estuary, and from the estuary they are swept into the sea. The coarser and heavier fragments are obviously deposited first, that is, as soon as the current begins to lose its force by becoming amalgamated with the stiller depths of the ocean, but the finer and lighter particles are carried further on, and eventually deposited in a deeper and stiller portion of the ocean.

It clearly follows from this that mud gives us a chronology; for it is evident that supposing this, which I now sketch, to be the sea bottom, and supposing this to be a coast-line; from the washing action of the sea upon the rock, wearing and grinding it down into a sediment of mud, the mud will be carried down, and at length, deposited in the deeper parts of this sea bottom, where it will form a layer; and then, while that first layer is hardening, other mud which is coming from the same source will, of course, be carried to the same place; and, as it is quite impossible for it to get beneath the layer already there, it deposits itself above it, and forms another layer, and in that way you gradually have layers of mud constantly forming and hardening one above the other, and conveying a record of time.

It is a necessary result of the operation of the law of gravitation that the uppermost layer shall be the youngest and the lowest the oldest, and that the different beds shall be older at any particular point or spot in exactly the ratio of their depth from the surface. So that if they were upheaved afterwards, and you had a series of these different layers of mud, converted into sandstone, or limestone, as the case might be, you might be sure that the bottom layer was deposited first, and that the upper layers were formed afterwards. Here, you see, is the first step in the history—these layers of mud give us an idea of time.

The whole surface of the earth,—I speak broadly, and leave out minor qualifications,—is made up of such layers of mud, so hard, the majority of them, that we call them rock, whether limestone or sandstone, or other varieties of rock. And, seeing that every part of the crust of the earth is made up in this way, you might think that the determination of the chronology, the fixing of the time which it has taken to form this crust is a comparatively simple matter. Take a broad average, ascertain how fast the mud is deposited upon the bottom of the sea, or in the estuary of rivers; take it to be an inch, or two, or three inches a year, or whatever you may roughly estimate it at; then take the total thickness of the whole series of stratified rocks, which geologists estimate at twelve or thirteen miles, or about seventy thousand feet, make a sum in short division, divide the total thickness by that of the quantity deposited in one year, and the result will, of course, give you the number of years which the crust has taken to form.

Truly, that looks a very simple process! It would be so except for certain difficulties, the very first of which is that of finding how rapidly sediments are deposited; but the main difficulty—a difficulty which renders any certain calculations of such a matter out of the question—is this, the sea-bottom on which the deposit takes place is continually shifting.

Instead of the surface of the earth being that stable, fixed thing that it is popularly believed to be, being, in common parlance, the very emblem of fixity itself, it is incessantly moving, and is, in fact, as unstable as the surface of the sea, except that its undulations are infinitely slower and enormously higher and deeper.

Now, what is the effect of this oscillation? Take the case to which I have previously referred. The finer or coarser sediments that are carried down by the current of the river, will only be carried out a certain distance, and eventually, as we have already seen, on reaching the stiller part of the ocean, will be deposited at the bottom.

Let C y (Fig. 4) be the sea-bottom, y D the shore, x y the sea-level, then the coarser deposit will subside

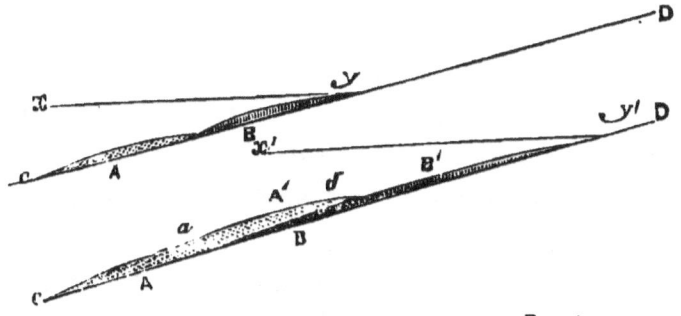

Fig. 4.

over the region B, the finer over A, while beyond A there will be no deposit at all; and, consequently, no record will be kept, simply because no deposit is going on. Now, suppose that the whole land, C, D, which we have regarded as stationary, goes down, as it does so, both A and B go further out from the shore, which will be at y^1, $x^1 y^1$, being the new sea-level. The consequence will be that the layer of mud (A), being now, for the most part, further than the force of the current is strong enough to convey even the finest *débris*, will, of course, receive no more deposits, and having attained a certain thickness will now grow no thicker.

We should be misled in taking the thickness of that layer, whenever it may be exposed to our view, as a record of time in the manner in which we are now regarding this subject, as it would give us only an imperfect and partial record: it would seem to represent too short a period of time.

Suppose, on the other hand, that the land (C D) had gone on rising slowly and gradually—say an inch or two inches in the course of a century,—

what would be the practical effect of that movement? Why, that the sediment A and B which has been already deposited, would eventually be brought nearer to the shore-level, and again subjected to the wear and tear of the sea; and directly the sea begins to act upon it, it would of course soon cut up and carry it away, to a greater or less extent, to be re-deposited further out.

Well, as there is, in all probability, not one single spot on the whole surface of the earth, which has not been up and down in this way a great many times, it follows that the thickness of the deposits formed at any particular spot cannot be taken (even supposing we had at first obtained correct data as to the rate at which they took place), as affording reliable information as to the period of time occupied in its deposit. So that you see it is absolutely necessary from these facts, seeing that our record entirely consists of accumulations of mud, superimposed one on the other; seeing in the next place that any particular spots on which accumulations have occurred, have been constantly moving up and down, and sometimes out of the reach of a deposit, and at other times its own deposit broken up and carried away, it follows that our record must be in the highest degree imperfect, and we have hardly a trace left of thick deposits, or any definite knowledge of the area that they occupied in a great many cases. And mark this! That supposing even that the whole surface of the earth had been accessible to the geologist,—that man had had access to every part of the earth, and had made sections of the whole, and put them all together,— even then his record must of necessity be imperfect.

But to how much has man really access? If you will look at this Map you will see that it represents the proportion of the sea to the earth: this coloured part indicates all the dry land, and this other portion is the water. You will notice at once that the water covers three-fifths of the whole surface of the globe, and has covered it in the same manner ever since man has kept any record of his own observations, to say nothing of the minute period during which he has cultivated geological inquiry. So that three-fifths of the surface of the earth is shut out from us because it is under the sea. Let us look at the other two-fifths, and see what are the countries in which anything that may be termed searching geological inquiry has been carried out: a good deal of France, Germany, and Great Britain and Ireland, bits of Spain, of Italy, and of Russia, have been examined, but of the whole great mass of Africa, except parts of the southern extremity, we know next to nothing; little bits of India, but of the greater part of the Asiatic continent nothing; bits of the Northern American States and of Canada, but of the greater part of the continent of North America, and in still larger proportion, of South America, nothing!

Under these circumstances, it follows that even with reference to that kind of imperfect information which we can possess, it is only of about the ten-thousandth part of the accessible parts of the earth that has been examined properly. Therefore, it is with justice that the most thoughtful of those who are concerned in these inquiries insist continually upon the imperfection of the geological record; for, I repeat, it is absolutely necessary, from the nature of things, that that record should be of the most fragmentary and imperfect character. Unfortunately this circumstance has been constantly forgotten. Men of science, like young colts in a fresh pasture, are apt to be exhilarated on being turned into a new field of inquiry, to go off at a hand-gallop, in total disregard of hedges and ditches, to lose sight of the real limitation of their inquiries, and to forget the extreme imperfection of what is really known. Geologists have imagined that they could tell us what was going on at all parts of the earth's surface during a given epoch; they have talked of this deposit being contemporaneous with that deposit, until, from our little local histories of the changes at limited spots of the earth's surface, they have constructed a universal history of the globe as full of wonders and portents as any other story of antiquity.

But what does this attempt to construct a universal history of the globe imply? It implies that we shall not only have a precise knowledge of the events which have occurred at any particular point, but that we shall be able to say what events, at any one spot, took place at the same time with those at other spots.

Let us see how far that is in the nature of things practicable. Suppose that here I make a section of the Lake of Killarney, and here the section of another lake—that of Loch Lomond in Scotland for instance. The rivers that flow into them are constantly carrying down deposits of mud, and beds, or strata, are being as constantly formed, one above the other, at the bottom of those lakes. Now, there is not a shadow of doubt that in these two lakes the lower beds are all older than the upper—there is no doubt about that; but what does *this* tell us about the age of any given bed in Loch Lomond, as compared with that of any given bed in the Lake of Killarney? It is, indeed, obvious that if any two sets of deposits are separated and discontinuous, there is absolutely no means whatever given you by the nature of the deposit of saying whether one is much younger or older than the other; but you may say, as many have said and think, that the case is very much altered if the beds which we are comparing are continuous. Suppose two beds of mud hardened into rock,—A and B are seen in section. (Fig. 5.)

Well, you say, it is admitted that the lowermost bed is always the older. Very well; B, therefore, is older than A. No doubt, *as a whole*, it

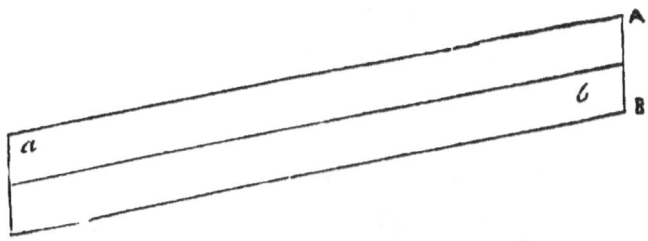

Fig. 5.

is so; or if any parts of the two beds which are in the same vertical line are compared, it is so. But suppose you take what seems a very natural step further, and say that the part *a* of the bed A is younger than the part *b* of the bed B. Is this sound reasoning? If you find any record of changes taking place at *b*, did they occur before any events which took place while *a* was being deposited? It looks all very plain sailing, indeed, to say that they did; and yet there is no proof of anything of the kind. As the former Director of this Institution, Sir H. De la Beche, long ago showed, this reasoning may involve an entire fallacy. It is extremely possible that *a* may have been deposited ages before *b*. It is very easy to understand how that can be. To return to Fig. 4; when A and B were deposited, they were *substantially* contemporaneous; A being simply the finer deposit, and B the coarser of the same detritus or waste of land. Now suppose that that sea-bottom goes down (as shown in Fig. 4), so that the first deposit is carried no farther than a, forming the bed A^1, and the coarse no farther than b, forming the bed B^1, the result will be the formation of two continuous beds, one of fine sediment ($A A^1$) over-lapping another of coarse sediment ($B B^1$). Now suppose the whole sea-bottom is raised up, and a section exposed about the point A^1; no doubt, *at this spot*, the upper bed is younger than the lower. But we should obviously greatly err if we concluded that the mass of the upper bed at A was younger than the lower bed at B; for we have just seen that they are contemporaneous deposits. Still more should we be in error if we supposed the upper bed at A to be younger than the continuation of the lower bed at B^1; for A was deposited long before B^1. In fine, if, instead of comparing immediately adjacent parts of two beds, one of which lies upon another, we compare distant parts, it is quite possible that the upper may be any number of years older than the under, and the under any number of years younger than the upper.

Now you must not suppose that I put this before

you for the purpose of raising a paradoxical difficulty; the fact is, that the great mass of deposits have taken place in sea-bottoms which are gradually sinking, and have been formed under the very conditions I am here supposing.

Do not run away with the notion that this subverts the principle I laid down at first. The error lies in extending a principle which is perfectly applicable to deposits in the same vertical line to deposits which are not in that relation to one another.

It is in consequence of circumstances of this kind, and of others that I might mention to you, that our conclusions on and interpretations of the record are really and strictly only valid so long as we confine ourselves to one vertical section. I do not mean to tell you that there are no qualifying circumstances, so that, even in very considerable areas, we may safely speak of conformably superimposed beds being older or younger than others at many different points. But we can never be quite sure in coming to that conclusion, and especially we cannot be sure if there is any break in their continuity, or any very great distance between the points to be compared.

Well now, so much for the record itself,—so much for its imperfections,—so much for the conditions to be observed in interpreting it, and its chronological indications, the moment we pass beyond the limits of a vertical linear section.

Now let us pass from the record to that which it contains,—from the book itself to the writing and the figures on its pages. This writing and these figures consist of remains of animals and plants which, in the great majority of cases, have lived and died in the very spot in which we now find them, or at least in the immediate vicinity. You must all of you be aware— and I referred to the fact in my last lecture—that there are vast numbers of creatures living at the bottom of the sea. These creatures, like all others, sooner or later die, and their shells and hard parts lie at the bottom; and then the fine mud which is being constantly brought down by rivers and the action of the wear and tear of the sea, covers them over and protects them from any further change or alteration; and, of course, as in process of time the mud becomes hardened and solidified, the shells of these animals are preserved and firmly embedded in the limestone or sandstone which is being thus formed. You may see in the galleries of the Museum upstairs specimens of limestones in which such fossil remains of existing animals are embedded. There are some specimens in which turtles' eggs have been imbedded in calcareous sand, and before the sun had hatched the young turtles, they became covered over with calcareous mud, and thus have been preserved and fossilized.

Not only does this process of imbedding and fossilization occur with marine and other aquatic animals and plants, but it affects those land animals and plants which are drifted away to sea, or become buried in bogs or morasses; and the animals which have been trodden down by their fellows and crushed in the mud at the river's bank, as the herd have come to drink. In any of these cases, the organisms may be crushed or be mutilated, before or after putrefaction, in such a manner that perhaps only a part will be left in the form in which it reaches us. It is, indeed, a most remarkable fact, that it is quite an exceptional case to find a skeleton of any one of all the thousands of wild land animals that we know are constantly being killed, or dying in the course of nature: they are preyed on and devoured by other animals, or die in places where their bodies are not afterwards protected by mud. There are other animals existing in the sea, the shells of which form exceedingly large deposits. You are probably aware that before the attempt was made to lay the Atlantic telegraphic cable, the Government employed vessels in making a series of very careful observations and soundings of the bottom of the Atlantic; and although, as we must all regret, that up to the present time that project has not succeeded, we have the satisfaction of knowing that it yielded some most remarkable results to science. The Atlantic Ocean had to be sounded right across, to depths of several miles in some places, and the nature of its bottom was carefully ascertained. Well, now, a space of about 1,000 miles wide from east to west, and I do not exactly know how many from north to south, but at any rate 600 or 700 miles, was carefully examined, and it was found that over the whole of that immense area an excessively fine chalky mud is being deposited; and this deposit is entirely made up of animals whose hard parts are deposited in this part of the ocean, and are doubtless gradually acquiring solidity and becoming metamorphosed into a chalky limestone. Thus, you see, it is quite possible in this way to preserve unmistakable records of animal and vegetable life. Whenever the sea-bottom, by some of those undulations of the earth's crust that I have referred to, becomes upheaved, and sections or borings are made, or pits are dug, then we become able to examine the contents and constituents of these ancient sea-bottoms, and find out what manner of animals lived at that period.

Now it is a very important consideration in its bearing on the completeness of the record, to inquire how far the remains contained in these fossiliferous limestones are able to convey anything like an accurate or complete account of the animals which were in existence at the time of its formation. Upon that point we can form a very clear judgment, and one in which there is no possible room for any mistake. There are of course a great number of animals—such as jelly-fishes, and other animals—without any hard parts, of which we cannot reasonably expect to find any traces whatever: there is nothing of them to preserve. Within a very short time, you will have noticed, after they are removed from the water, they dry up to a mere nothing; certainly they are not of a nature to leave any very visible traces of their existence on such bodies as chalk or mud. Then again, look at land animals; it is, as I have said, a very uncommon thing to find a land animal entire after death. Insects and other carnivorous animals very speedily pull them to pieces, putrefaction takes place, and so, out of the hundreds of thousands that are known to die every year, it is the rarest thing in the world to see one imbedded in such a way that its remains would be preserved for a lengthened period. Not only is this the case, but even when animal remains have been safely imbedded, certain natural agents may wholly destroy and remove them.

Almost all the hard parts of animals—the bones and so on—are composed chiefly of phosphate of lime and carbonate of lime. Some years ago, I had to make an inquiry into the nature of some very curious fossils sent to me from the North of Scotland. Fossils are usually hard bony structures that have become imbedded in the way I have described, and have gradually acquired the nature and solidity of the body with which they are associated; but in this case I had a series of *holes* in some pieces of rock, and nothing else. Those holes, however, had a certain definite shape about them, and when I got a skilful workman to make castings of the interior of these holes, I found that they were the impressions of the joints of a back-bone and of the armour of a great reptile, twelve or more feet long. This great beast had died and got buried in the sand, the sand had gradually hardened over the bones, but remained porous. Water had trickled through it, and that water being probably charged with a superfluity of carbonic acid, had dissolved all the phosphate and carbonate of lime, and the bones themselves had thus decayed and entirely disappeared; but as the sandstone happened to have consolidated by that time, the precise shape of the bones was retained. If that sandstone had remained soft a little longer, we should have known nothing whatsoever of the existence of the reptile whose bones it had encased.

How certain it is that a vast number of animals which have existed at one period on this earth have entirely perished, and left no trace whatever of their forms, may be proved to you by other considerations. There are large tracts of sandstone in various parts of the world, in which nobody has yet found anything but footsteps. Not a bone of any description, but an enormous number of traces of footsteps. There is no question about them. There is a whole valley in Connecticut covered with these footsteps, and not a single fragment of the animals which made them have yet been found. Let me mention another case while upon that matter, which is even more surprising than those to which I have yet referred. There is a limestone formation near Oxford, at a place called Stonesfield, which has yielded the remains of certain very interesting mammalian animals, and up to this time, if I recollect rightly, there have been found seven specimens of its lower jaws, and not a bit of anything else, neither limb-bones nor skull, or any part whatever; not a fragment of the whole system! Of course, it would be preposterous to imagine that the beasts had nothing else but a lower jaw! The probability is, as Dr. Buckland showed, as the result of his observations on dead dogs in the river Thames, that the lower jaw, not being secured by very firm ligaments to the bones of the head, and being a weighty affair, would easily be knocked off, or might drop away from the body as it floated in water in a state of decomposition. The jaw would thus be deposited immediately, while the rest of the body would float and drift away altogether, ultimately reaching the sea, and perhaps becoming destroyed. The jaw becomes covered up and preserved in the river silt, and thus it comes that we have such a curious circumstance as that of the lower jaws in the Stonesfield slates. So that, you see, faulty as these layers of stone in the earth's crust are, defective as they necessarily are as a record, the account of contemporaneous vital phenomena presented by them is, by the necessity of the case, infinitely more defective and fragmentary.

It was necessary that I should put all this very strongly before you, because, otherwise, you might have been led to think differently of the completeness of our knowledge by the next facts I shall state to you.

The researches of the last three-quarters of a century have, in truth, revealed a wonderful richness of organic life in those rocks. Certainly not fewer than thirty or forty thousand different species of fossils have been discovered. You have no more ground for doubting that these creatures really lived and died at or near the places in which we find them than you have for like scepticism about a shell on the sea-shore. The evidence is as good in the one case as in the other.

Our next business is to look at the general character of these fossil remains, and it is a subject which will be requisite to consider carefully; and the first point for us is to examine how much the extinct *Flora* and *Fauna* as a *whole*—disregarding altogether the *succession* of their constituents, of which I shall speak afterwards—differ from the *Flora* and *Fauna* of the present day;—how far they differ in what we *do* know about them, leaving altogether out of consideration speculations based on what we *do not* know.

I strongly imagine that if it were not for the peculiar appearance that fossilized animals have, that any of you might readily walk through a museum which contains fossil remains mixed up with those of the present forms of life, and I doubt very much whether your uninstructed eyes would lead you to see any vast or wonderful difference between the two. If you looked closely, you would notice, in the first place, a great many things very like animals with which you are acquainted now: you would see differences of shape and proportion, but on the whole a close similarity.

I explained what I meant by ORDERS the other day, when I described the animal kingdom as being divided into sub-kingdoms, classes, and orders. If you divide the animal kingdom into orders, you will find that there are above one hundred and twenty. The number may vary on one side or the other, but this is a fair estimate. That is the sum total of the orders of all the animals which we know now, and which have been known in past times, and left remains behind.

Now, how many of those are absolutely extinct? That is to say, how many of these orders of animals have lived at a former period of the world's history, but have at present no representatives? That is the sense in which I meant to use the word "extinct." I mean that those animals did live on this earth at one time, but have left no one of their kind with us at the present moment. So that estimating the number of extinct animals is a sort of way of comparing the past creation as a whole with the present as a whole. Among the mammalia and birds there are none extinct; but when we come to the reptiles there is a most wonderful thing: out of the eight orders, or thereabouts, which you can make among reptiles, one-half are extinct. These diagrams of the plesiosaurus, the ichthyosaurus, the pterodactyle, give you a notion of some of these extinct reptiles. And here is a cast of the pterodactyle and bones of the ichthyosaurus and the plesiosaurus, just as fresh as if it had been recently dug up in a churchyard. Thus, in the reptile class, there are no less than half of the orders which are absolutely extinct. If we turn to the *Amphibia*, there was one extinct order, the Labyrinthodonts, typified by the large salamander-like beast shown in this diagram.

No order of fishes is known to be extinct. Every fish that we find in the strata—to which I have been referring—can be identified and placed in one of the orders which exist at the present day. There is not known to be a single ordinal form of insect extinct. There are only two orders extinct among the *Crustacea*. There is not known to be an extinct order of these creatures, the parasitic and other worms; but there are two, not to say three, absolutely extinct orders of this class, the *Echinodermata*; out of all the orders of the *Cœlenterata* and *Protozoa* only one, the Rugose Corals.

So that, you see, out of somewhere about 120 orders of animals, taking them altogether, you will not, at the outside estimate, find above ten or a dozen extinct. Summing up all the order of animals which have left remains behind them, you will not find above ten or a dozen which cannot be arranged with those of the present day; that is to say, that the difference does not amount to much more than ten per cent.: and the

D

proportion of extinct orders of plants is still smaller. I think that that is a very astounding, a most astonishing fact: seeing the enormous epochs of time which have elapsed during the constitution of the surface of the earth as it at present exists; it is, indeed, a most astounding thing that the proportion of extinct ordinal types should be so exceedingly small.

But now, there is another point of view in which we must look at this past creation. Suppose that we were to sink a vertical pit through the floor beneath us, and that I could succeed in making a section right through in the direction of New Zealand, I should find in each of the different beds through which I passed the remains of animals which I should find in that stratum and not in the others. First, I should come upon beds of gravel or drift containing the bones of large animals, such as the elephant, rhinoceros, and cave tiger. Rather curious things to fall across in Piccadilly! If I should dig lower still, I should come upon a bed of what we call the London clay, and in this, as you will see in our galleries up-stairs, are found remains of strange cattle, remains of turtles, palms, and large tropical fruits; with shell-fish such as you see the like of now only in tropical regions. If I went below that, I should come upon the chalk, and there I should find something altogether different, the remains of ichthyosauri and pterodactyles, and ammonites, and so forth.

I do not know what Mr. Godwin Austin would say comes next, but probably rocks containing more ammonites, and more ichthyosauri and plesiosauri, with a vast number of other things; and under that I should meet with yet older rocks, containing numbers of strange shells and fishes; and in thus passing from the surface to the lowest depths of the earth's crust, the forms of animal life and vegetable life which I should meet with in the successive beds would, looking at them broadly, be the more different the further that I went down. Or, in other words, inasmuch as we started with the clear principle, that in a series of naturally-disposed mud beds the lowest are the oldest, we should come to this result, that the further we go back in time the more difference exists between the animal and vegetable life of an epoch and that which now exists. That was the conclusion to which I wished to bring you at the end of this Lecture.

LECTURE III.

THE METHOD BY WHICH THE CAUSES OF THE PRESENT AND PAST CONDITIONS OF ORGANIC NATURE ARE TO BE DISCOVERED. — THE ORIGINATION OF LIVING BEINGS.

IN the two preceding lectures I have endeavoured to indicate to you the extent of the subject-matter of the inquiry upon which we are engaged; and having thus acquired some conception of the Past and Present phenomena of Organic Nature, I must now turn to that which constitutes the great problem which we have set before ourselves;—I mean, the question of what knowledge we have of the causes of these phenomena of organic nature, and how such knowledge is obtainable.

Here, on the threshold of the inquiry, an objection meets us. There are in the world a number of extremely worthy, well-meaning persons, whose judgments and opinions are entitled to the utmost respect on account of their sincerity, who are of opinion that Vital Phenomena, and especially all questions relating to the origin of vital phenomena, are questions quite apart from the ordinary run of inquiry, and are, by their very nature, placed out of our reach. They say that all these phenomena originated miraculously, or in some way totally different from the ordinary course of nature, and that therefore they conceive it to be futile, not to say presumptuous, to attempt to inquire into them.

To such sincere and earnest persons, I would only say, that a question of this kind is not to be shelved upon theoretical or speculative grounds. You may remember the story of the Sophist who demonstrated to Diogenes in the most complete and satisfactory manner that he could not walk; that, in fact, all motion was an impossibility; and that Diogenes refuted him by simply getting up and walking round his tub. So, in the same way, the man of science replies to objections of this kind, by simply getting up and walking onward, and showing what science has done and is doing,—by pointing to that immense mass of facts which have been ascertained and systematized under the forms of the great doctrines of Morphology, of Development, of Distribution, and the like. He sees an enormous mass of facts and laws relating to organic beings, which stand on the same good sound foundation as every other natural law. With this mass of facts and laws before us, therefore, seeing that, as far as organic matters have hitherto been accessible and studied, they have shown themselves capable of yielding to scientific investigation, we may accept this as proof that order and law reign there as well as in the rest of nature. The man of science says nothing to objectors of this sort, but supposes that we can and shall walk to a knowledge of the origin of organic nature, in the same way that we have walked to a knowledge of the laws and principles of the inorganic world.

But there are objectors who say the same from ignorance and ill-will. To such I would reply that the objection comes ill from them, and that the real presumption, I may almost say the real blasphemy, in this matter, is in the attempt to limit that inquiry into the causes of phenomena, which is the source of all human blessings, and from which has sprung all human prosperity and progress; for, after all, we can accomplish comparatively little; the limited range of our own faculties bounds us on every side,—the field of our powers of observation is small enough, and he who endeavours to narrow the sphere of our inquiries is only pursuing a course that is likely to produce the greatest harm to his fellow-men.

But now, assuming, as we all do, I hope, that these phenomena are properly accessible to inquiry, and setting out upon our search into the causes of the phenomena of organic nature, or, at any rate, setting out to discover how much we at present know upon these abstruse matters, the question arises as to what is to be our course of proceeding, and what method we must lay down for our guidance. I reply to that question, that our method must be exactly the same as that which is pursued in any other scientific inquiry, the method of scientific investigation being the same for all orders of facts and phenomena whatsoever.

I must dwell a little on this point, for I wish you to

leave this room with a very clear conviction that scientific investigation is not, as many people seem to suppose, some kind of modern black art. I say that you might easily gather this impression from the manner in which many persons speak of scientific inquiry, or talk about inductive and deductive philosophy, or the principles of the " Baconian philosophy." I do protest that, of the vast number of cants in this world, there are none, to my mind, so contemptible as the pseudo-scientific cant which is talked about the "Baconian philosophy."

To hear people talk about the great Chancellor,—and a very great man he certainly was,—you would think that it was he who had invented science, and that there was no such thing as sound reasoning before the time of Queen Elizabeth! Of course you say, that cannot possibly be true; you perceive, on a moment's reflection, that such an idea is absurdly wrong; and yet, so firmly rooted is this sort of impression,—I cannot call it an idea, or conception,—the thing is too absurd to be entertained,—but so completely does it exist at the bottom of most men's minds, that this has been a matter of observation with me for many years past. There are many men who, though knowing absolutely nothing of the subject with which they may be dealing, wish, nevertheless, to damage the author of some view with which they think fit to disagree. What they do, then, is not to go and learn something about the subject, which one would naturally think the best way of fairly dealing with it; but they abuse the originator of the view they question, in a general manner, and wind up by saying that, "After all, you know, the principles and method of this author are totally opposed to the canons of the Baconian philosophy." Then everybody applauds, as a matter of course, and agrees that it must be so. But if you were to stop them all in the middle of their applause, you would probably find that neither the speaker nor his applauders could tell you how or in what way it was so; neither the one nor the other having the slightest idea of what they mean when they speak of the "Baconian philosophy."

You will understand, I hope, that I have not the slightest desire to join in the outcry against either the morals, the intellect, or the great genius of Lord Chancellor Bacon. He was undoubtedly a very great man, let people say what they will of him; but notwithstanding all that he did for philosophy, it would be entirely wrong to suppose that the methods of modern scientific inquiry originated with him, or with his age; they originated with the first man, whoever he was; and indeed existed long before him, for many of the essential processes of reasoning are exerted by the higher order of brutes as completely and effectively as by ourselves. We see in many of the brute creation the exercise of one, at least, of the same powers of reasoning as that which we ourselves employ.

The method of scientific investigation is nothing but the expression of the necessary mode of working of the human mind. It is simply the mode at which all phenomena are reasoned about, rendered precise and exact. There is no more difference, but there is just the same kind of difference, between the mental operations of a man of science and those of an ordinary person, as there is between the operations and methods of a baker or of a butcher weighing out his goods in common scales, and the operations of a chemist in performing a difficult and complex analysis by means of his balance and finely-graduated weights. It is not that the action of the scales in the one case, and the balance in the other, differ in the principles of their construction or manner of working; but the beam of one is set on an infinitely finer axis than the other, and of course turns by the addition of a much smaller weight.

You will understand this better, perhaps, if I give you some familiar example. You have all heard it repeated, I dare say, that men of science work by means of Induction and Deduction, and that by the help of these operations, they, in a sort of sense, wring from Nature certain other things, which are called Natural Laws, and Causes, and that out of these, by some cunning skill of their own, they build up Hypotheses and Theories. And it is imagined by many, that the operations of the common mind can be by no means compared with these processes, and that they have to be acquired by a sort of special apprenticeship to the craft. To hear all these large words, you would think that the mind of a man of science must be constituted differently from that of his fellow men; but if you will not be frightened by terms, you will discover that you are quite wrong, and that all these terrible apparatus are being used by yourselves every day and every hour of your lives.

There is a well-known incident in one of Molière's plays, where the author makes the hero express unbounded delight on being told that he had been talking prose during the whole of his life. In the same way, I trust, that you will take comfort, and be delighted with yourselves, on the discovery that you have been acting on the principles of inductive and deductive philosophy during the same period. Probably there is not one here who has not in the course of the day had occasion to set in motion a complex train of reasoning, of the very same kind, though differing of course in degree, as that which a scientific man goes through in tracing the causes of natural phenomena.

A very trivial circumstance will serve to exemplify this. Suppose you go into a fruiterer's shop, wanting an apple,—you take up one, and, on biting it, you find it is sour; you look at it, and see that it is hard and green. You take up another one, and that too is hard, green, and sour. The shopman offers you a third; but, before biting it, you examine it, and find that it is hard and green, and you immediately say that you will not have it, as it must be sour, like those that you have already tried.

Nothing can be more simple than that, you think; but if you will take the trouble to analyze and trace out into its logical elements what has been done by the mind, you will be greatly surprised. In the first place, you have performed the operation of Induction. You found that, in two experiences, hardness and greenness in apples went together with sourness.

It was so in the first case, and it was confirmed by the second. True, it is a very small basis, but still it is enough to make an induction from; you generalize the facts, and you expect to find sourness in apples where you get hardness and greenness. You found upon that a general law, that all hard and green apples are sour; and that, so far as it goes, is a perfect induction. Well, having got your natural law in this way, when you are offered another apple which you find is hard and green, you say, "All hard and green apples are sour; this apple is hard and green, therefore this apple is sour." That train of reasoning is what logicians call a syllogism, and has all its various parts and terms,—its major premiss, its minor premiss, and its conclusion. And, by the help of further reasoning, which, if drawn out, would have to be exhibited in two or three other syllogisms, you arrive at your final determination, "I will not have that apple." So that, you see, you have, in the first place, established a law by Induction, and upon that you have founded a Deduction, and reasoned out the special conclusion of the particular case. Well now, suppose, having got your law, that at some time afterwards, you are discussing the qualities of apples with a friend: you will say to him, "It is a very curious thing,—but I find that all hard and green apples are sour!" Your friend says to you, "But how do you know that?" You at once reply, "Oh, because I have tried them over and over again, and have always found them to be so." Well, if we were talking science instead of common sense, we should call that an Experimental Verification.

And, if still opposed, you go further, and say, "I have heard from the people in Somersetshire and Devonshire, where a large number of apples are grown, that they have observed the same thing. It is also found to be the case in Normandy, and in North America. In short, I find it to be the universal experience of mankind wherever attention has been directed to the subject." Whereupon, your friend, unless he is a very unreasonable man, agrees with you, and is convinced that you are quite right in the conclusion you have drawn. He believes, although perhaps he does not know he believes it, that the more extensive Verifications are,—that the more frequently experiments have been made, and results of the same kind arrived at,—that the more varied the conditions under which the same results are attained, the more certain is the ultimate conclusion, and he disputes the question no further. He sees that the experiment has been tried under all sorts of conditions, as to time, place, and people, with the same result; and he says with you, therefore, that the law you have laid down must be a good one, and he must believe it.

In science we do the same thing;—the philosopher exercises precisely the same faculties, though in a much more delicate manner. In scientific inquiry it becomes a matter of duty to expose a supposed law to every possible kind of verification, and to take care, moreover, that this is done intentionally, and not left to a mere accident, as in the case of the apples. And in science, as in common life, our confidence in a law is in exact proportion to the absence of variation in the result of our experimental verifications. For instance, if you let go your grasp of an article you may have in your hand, it will immediately fall to the ground. That is a very common verification of one of the best established laws of nature—that of gravitation. The method by which men of science establish the existence of that law is exactly the same as that by which we have established the trivial proposition about the sourness of hard and green apples. But we believe it in such an extensive, thorough, and unhesitating manner because the universal experience of mankind verifies it, and we can verify it ourselves at any time; and that is the strongest possible foundation on which any natural law can rest.

So much, then, by way of proof that the method of establishing laws in science is exactly the same as that pursued in common life. Let us now turn to another matter, (though really it is but another phase of the same question,) and that is, the method by which, from the relations of certain phenomena, we prove that some stand in the position of causes towards the others.

I want to put the case clearly before you, and I will therefore show you what I mean by another familiar example. I will suppose that one of you, on coming down in the morning to the parlour of your house, finds that a tea-pot and some spoons which had been left in the room on the previous evening are gone, —the window is open, and you observe the mark of a dirty hand on the window-frame, and perhaps, in addition to that, you notice the impress of a hob-nailed shoe on the gravel outside. All these phenomena have struck your attention instantly, and before two seconds have passed you say, "Oh, somebody has broken open the window, entered the room, and run off with the spoons and the tea-pot!" That speech is out of your mouth in a moment. And you will probably add, "I know there has; I am quite sure of it!" You mean to say exactly what you know; but in reality you are giving expression to what is, in all essential particulars, an Hypothesis. You do not *know* it at all; it is nothing but an hypothesis rapidly framed in your own mind! And, it is an hypothesis founded on a long train of inductions and deductions.

What are those inductions and deductions, and how have you got at this hypothesis? You have observed, in the first place, that the window is open; but by a train of reasoning involving many Inductions and Deductions, you have probably arrived long before at the General Law — and a very good one it is — that windows do not open of themselves; and you therefore conclude that something has opened the window. A second general law that you have arrived at in the same way is, that tea-pots and spoons do not go out of a window spontaneously, and you are satisfied that, as they are not now where you left them, they have been removed. In the third place, you look at the marks on the window-sill, and the shoe-marks outside, and you say that in all previous experience the former kind of mark has never been produced by anything else but the hand of a human being; and the same experience shows that no other animal but man at present wears shoes with hob-nails in them such as would produce the marks in the gravel. I do not know, even if we could discover any of those "missing links" that are talked about, that they would help us to any other conclusion! At any rate the law which states our present experience is strong enough for my present purpose. You next reach the conclusion, that as these kinds of marks have not been left by any other animals than men, or are liable to be formed in any other way than by a man's hand and shoe, the marks in question have been formed by a man in that way. You have, further, a general law, founded on observation and experience, and that, too, is, I am sorry to say, a very universal and unimpeachable one,—that some men are thieves; and you assume at once from all these premisses—and that is what constitutes your hypothesis—that the man who made the marks outside and on the window-sill, opened the window, got into the room, and stole your tea-pot and spoons. You have now arrived at a *Vera Causa*;—you have assumed a Cause which it is plain is competent to produce all the phenomena you have observed. You can explain all these phenomena only by the hypothesis of a thief. But that is a hypothetical conclusion, of the justice of which you have no absolute proof at all; it is only rendered highly probable by a series of inductive and deductive reasonings.

I suppose your first action, assuming that you are a man of ordinary common sense, and that you have established this hypothesis to your own satisfaction, will very likely be to go off for the police, and set them on the track of the burglar, with the view to the recovery of your property. But just as you are starting with this object, some person comes in, and on learning what you are about, says, " My good friend, you are going on a great deal too fast. How do you know that the man who really made the marks took the spoons? It might have been a monkey that took them, and the man may have merely looked in afterwards." You ould probably reply, " Well, that is all very well, but you see it is contrary to all experience of the way tea-pots and spoons are abstracted; so that, at any rate, your hypothesis is less probable than mine." While you are talking the thing over in this way, another friend arrives, one of that good kind of people that I was talking of a little while ago. And he might say, "Oh, my dear sir, you are certainly going on a great deal too fast. You are most presumptuous. You admit that all these occurrences took place when you were fast asleep, at a time when you could not possibly have known anything about what was taking place. How do you know that the laws of Nature are not suspended during the night? It may be that there has been some kind of supernatural interference in this case." In point of fact, he declares that your hypothesis is one of which you cannot at all demonstrate the truth, and that you are by no means sure that the laws of Nature are the same when you are asleep as when you are awake.

Well, now, you cannot at the moment answer that

kind of reasoning. You feel that your worthy friend has you somewhat at a disadvantage. You will feel perfectly convinced in your own mind, however, that you are quite right, and you say to him, "My good friend, I can only be guided by the natural probabilities of the case, and if you will be kind enough to stand aside and permit me to pass, I will go and fetch the police." Well, we will suppose that your journey is successful, and that by good luck you meet with a policeman; that eventually the burglar is found with your property on his person, and the marks correspond to his hand and to his boots. Probably any jury would consider those facts a very good experimental verification of your hypothesis, touching the cause of the abnormal phenomena observed in your parlour, and would act accordingly.

Now, in this supposititious case, I have taken phenomena of a very common kind, in order that you might see what are the different steps in an ordinary process of reasoning, if you will only take the trouble to analyze it carefully. All the operations I have described, you will see, are involved in the mind of any man of sense in leading him to a conclusion as to the course he should take in order to make good a robbery and punish the offender. I say that you are led, in that case, to your conclusion by exactly the same train of reasoning as that which a man of science pursues when he is endeavouring to discover the origin and laws of the most occult phenomena. The process is, and always must be, the same; and precisely the same mode of reasoning was employed by Newton and Laplace in their endeavours to discover and define the causes of the movements of the heavenly bodies, as you, with your own common sense, would employ to detect a burglar. The only difference is, that the nature of the inquiry being more abstruse, every step has to be most carefully watched, so that there may not be a single crack or flaw in your hypothesis. A flaw or crack in many of the hypotheses of daily life may be of little or no moment as affecting the general correctness of the conclusions at which we may arrive; but in a scientific inquiry a fallacy, great or small, is always of importance, and is sure to be in the long run constantly productive of mischievous, if not fatal results.

Do not allow yourselves to be misled by the common notion that an hypothesis is untrustworthy simply because it is an hypothesis. It is often urged, in respect to some scientific conclusion, that, after all, it is only an hypothesis. But what more have we to guide us in nine-tenths of the most important affairs of daily life than hypotheses, and often very ill-based ones? So that in science, where the evidence of an hypothesis is subjected to the most rigid examination, we may rightly pursue the same course. You may have hypotheses and hypotheses. A man may say, if he likes, that the moon is made of green cheese: that is an hypothesis. But another man, who has devoted a great deal of time and attention to the subject, and availed himself of the most powerful telescopes and the results of the observations of others, declares that in his opinion it is probably composed of materials very similar to those of which our own earth is made up: and that is also only an hypothesis. But I need not tell you that there is an enormous difference in the value of the two hypotheses. That one which is based on sound scientific knowledge is sure to have a corresponding value; and that which is a mere hasty random guess is likely to have but little value. Every great step in our progress in discovering causes has been made in exactly the same way as that which I have detailed to you. A person observing the occurrence of certain facts and phenomena asks, naturally enough, what process, what kind of operation known to occur in nature applied to the particular case, will unravel and explain the mystery? Hence you have the scientific hypothesis; and its value will be proportionate to the care and completeness with which its basis had been tested and verified. It is in these matters as in the commonest affairs of practical life: the guess of the fool will be folly, while the guess of the wise man will contain wisdom. In all cases, you see that the value of the result depends on the patience and faithfulness with which the investigator applies to his hypothesis every possible kind of verification.

I dare say I may have to return to this point by-and-by; but having dealt thus far with our logical methods, I must now turn to something which, perhaps, you may consider more interesting, or, at any rate, more tangible. But in reality there are but few things that can be more important for you to understand than the mental processes and the means by which we obtain scientific conclusions and theories.* Having granted that the inquiry is a proper one, and having determined on the nature of the methods we are to pursue and which only can lead to success, I must now turn to the consideration of our knowledge of the nature of the processes which have resulted in the present condition of organic nature.

Here, let me say at once, lest some of you misunderstand me, that I have extremely little to report. The question of how the present condition of organic nature came about, resolves itself into two questions. The first is: How has organic or living matter commenced its existence? And the second is: How has it been perpetuated? On the second question I shall have more to say hereafter. But on the first one, what I now have to say will be for the most part of a negative character.

If you consider what kind of evidence we can have upon this matter, it will resolve itself into two kinds. We may have historical evidence and we may have experimental evidence. It is, for example, conceivable, that inasmuch as the hardened mud which forms a considerable portion of the thickness of the earth's crust contains faithful records of the past forms of life, and inasmuch as these differ more and more as we go further down,—it is possible and conceivable that we might come to some particular bed or stratum which should contain the remains of those

---

* Those who wish to study fully the doctrines of which I have endeavoured to give some rough and ready illustrations, must read Mr. John Stuart Mill's "System of Logic."

creatures with which organic life began upon the earth. And if we did so, and if such forms of organic life were preservable, we should have what I would call historical evidence of the mode in which organic life began upon this planet. Many persons will tell you, and indeed you will find it stated in many works on geology, that this has been done, and that we really possess such a record; there are some who imagine that the earliest forms of life of which we have as yet discovered any record, are in truth the forms in which animal life began upon the globe. The grounds on which they base that supposition are these:—That if you go through the enormous thickness of the earth's crust and get down to the older rocks, the higher vertebrate animals—the quadrupeds, birds, and fishes—cease to be found; beneath them you find only the invertebrate animals; and in the deepest and lowest rocks those remains become scantier and scantier, not in any very gradual progression, however, until, at length, in what are supposed to be the oldest rocks, the animal remains which are found are almost always confined to four forms,— *Oldhamia*, whose precise nature is not known, whether plant or animal; *Lingula*, a kind of mollusc; *Trilobites*, a crustacean animal, having the same essential plan of construction, though differing in many details from a lobster or crab; and *Hymenocaris*, which is also a crustacean. So that you have all the *Fauna* reduced, at this period, to four forms: one a kind of animal or plant that we know nothing about, and three undoubted animals—two crustaceans and one mollusc.

I think, considering the organization of these mollusca and crustacea, and looking at their very complex nature, that it does indeed require a very strong imagination to conceive that these were the first created of all living things. And you must take into consideration the fact that we have not the slightest proof that these which we call the oldest beds are really so: I repeat, we have not the slightest proof of it. When you find in some places that in an enormous thickness of rocks there are but very scanty traces of life, or absolutely none at all; and that in other parts of the world rocks of the very same formation are crowded with the records of living forms, I think it is impossible to place any reliance on the supposition, or to feel oneself justified in supposing that these are the forms in which life first commenced. I have not time here to enter upon the technical grounds upon which I am led to this conclusion,—that could hardly be done properly in half a dozen lectures on that part alone;—I must content myself with saying that I do not at all believe that these are the oldest forms of life.

I turn to the experimental side to see what evidence we have there. To enable us to say that we know anything about the experimental origination of organization and life, the investigator ought to be able to take inorganic matters, such as carbonic acid, ammonia, water, and salines, in any sort of inorganic combination, and be able to build them up into Protein matter, and then that Protein matter ought to begin to live in an organic form. That, nobody has done as yet, and I suspect it will be a long while before anybody does do it. But the thing is by no means so impossible as it looks; for the researches of modern chemistry have shown us—I won't say the road towards it, but, if I may so say, they have shown the finger-post pointing to the road that may lead to it.

It is not many years ago—and you must recollect that Organic Chemistry is a young science, not above a couple of generations old, you must not expect too much of it,—it is not many years ago since it was said to be perfectly impossible to fabricate any organic compound; that is to say, any non-mineral compound which is to be found in an organized being. It remained so for a very long period; but it is now a considerable number of years since a distinguished foreign chemist contrived to fabricate Urea, a substance of a very complex character, which forms one of the waste products of animal structures. And of late years a number of other compounds, such as Butyric Acid, and others, have been added to the list. I need not tell you that chemistry is an enormous distance from the goal I indicate; all I wish to point out to you is, that it is by no means safe to say that that goal may not be reached one day. It may be that it is impossible for us to produce the conditions requisite to the origination of life; but we must speak modestly about the matter, and recollect that Science has put her foot upon the bottom round of the ladder. Truly he would be a bold man who would venture to predict where she will be fifty years hence.

There is another inquiry which bears indirectly upon this question, and upon which I must say a few words. You are all of you aware of the phenomena of what is called spontaneous generation. Our forefathers, down to the seventeenth century, or thereabouts, all imagined, in perfectly good faith, that certain vegetable and animal forms gave birth, in the process of their decomposition, to insect life. Thus, if you put a piece of meat in the sun, and allowed it to putrefy, they conceived that the grubs which soon began to appear were the result of the action of a power of spontaneous generation which the meat contained. And they could give you receipts for making various animal and vegetable preparations which would produce particular kinds of animals. A very distinguished Italian naturalist, named Redi, took up the question, at a time when everybody believed in it; among others our own great Harvey, the discoverer of the circulation of the blood. You will constantly find his name quoted, however, as an opponent of the doctrine of spontaneous generation; but the fact is, and you will see it if you will take the trouble to look into his works, Harvey believed it as profoundly as any man of his time; but he happened to enunciate a very curious proposition — that every living thing came from an *egg;* he did not mean to use the word in the sense in which we now employ it, he only meant to say that every living thing originated in a little rounded particle of organized substance; and it is from this circumstance, probably, that the notion of Harvey having opposed the doctrine originated. Then came Redi, and he proceeded to upset the doctrine in a very simple manner. He merely covered the piece of

ORIGINATION OF LIVING BEINGS. 73

meat with some very fine gauze, and then he exposed it to the same conditions. The result of this was that no grubs or insects were produced; he proved that the grubs originated from the insects who came and deposited their eggs in the meat, and that they were hatched by the heat of the sun. By this kind of inquiry he thoroughly upset the doctrine of spontaneous generation, for his time at least.

Then came the discovery and application of the microscope to scientific inquiries, which showed to naturalists that besides the organisms which they already knew as living beings and plants, there were an immense number of minute things which could be obtained apparently almost at will from decaying vegetable and animal forms. Thus, if you took some ordinary black pepper or some hay, and steeped it in water, you would find in the course of a few days that the water had become impregnated with an immense number of animalcules swimming about in all directions. From facts of this kind naturalists were led to revive the theory of spontaneous generation. They were headed here by an English naturalist,—Needham,—and afterwards in France by the learned Buffon. They said that these things were absolutely begotten in the water of the decaying substances out of which the infusion was made. It did not matter whether you took animal or vegetable matter, you had only to steep it in water and expose it, and you would soon have plenty of animalcules. They made an hypothesis about this which was a very fair one. They said, this matter of the animal world, or of the higher plants, appears to be dead, but in reality it has a sort of dim life about it, which, if it is placed under fair conditions, will cause it to break up into the forms of these little animalcules, and they will go through their lives in the same way as the animal or plant of which they once formed a part.

The question now became very hotly debated. Spallanzani, an Italian naturalist, took up opposite views to those of Needham and Buffon, and by means of certain experiments he showed that it was quite possible to stop the process by boiling the water, and closing the vessel in which it was contained. "Oh!" said his opponents; "but what do you know you may be doing when you heat the air over the water in this way? You may be destroying some property of the air requisite for the spontaneous generation of the animalcules."

However, Spallanzani's views were supposed to be upon the right side, and those of the others fell into discredit; although the fact was that Spallanzani had not made good his views. Well, then, the subject continued to be revived from time to time, and experiments were made by several persons; but these experiments were not altogether satisfactory. It was found that if you put an infusion in which animalcules would appear if it were exposed to the air into a vessel and boiled it, and then sealed up the mouth of the vessel, so that no air, save such as had been heated to 212°, could reach its contents, that then no animalcules would be found; but if you took the same vessel and exposed the infusion to the air, then you would get animalcules. Furthermore, it was found that if you connected the mouth of the vessel with a red-hot tube in such a way that the air would have to pass through the tube before reaching the infusion, that then you would get no animalcules. Yet another thing was noticed: if you took two flasks containing the same kind of infusion, and left one entirely exposed to the air, and in the mouth of the other placed a ball of cotton wool, so that the air would have to filter itself through it before reaching the infusion, that then, although you might have plenty of animalcules in the first flask, you would certainly obtain none from the second.

These experiments, you see, all tended towards one conclusion — that the infusoria were developed from little minute spores or eggs which were constantly floating in the atmosphere, and which lose their power of germination if subjected to heat. But one observer now made another experiment, which seemed to go entirely the other way, and puzzled him altogether. He took some of this boiled infusion that I have been speaking of, and by the use of a mercurial bath—a kind of trough used in laboratories—he deftly inverted a vessel containing the infusion into the mercury, so that the latter reached a little beyond the level of the mouth of the *inverted* vessel. You see that he thus had a quantity of the infusion shut off from any possible communication with the outer air by being inverted upon a bed of mercury.

He then prepared some pure oxygen and nitrogen gases, and passed them by means of a tube going from the outside of the vessel, up through the mercury into the infusion; so that he thus had it exposed to a perfectly pure atmosphere of the same constituents as the external air. Of course, he expected he would get no infusorial animalcules at all in that infusion; but, to his great dismay and discomfiture, he found he almost always did get them.

Furthermore, it has been found that experiments made in the manner described above answer well with most infusions; but that if you fill the vessel with boiled milk, and then stop the neck with cotton-wool, you *will* have infusoria. So that you see there were two experiments that brought you to one kind of conclusion, and three to another; which was a most unsatisfactory state of things to arrive at in a scientific inquiry.

Some few years after this, the question began to be very hotly discussed in France. There was M. Pouchet, a professor at Rouen, a very learned man, but certainly not a very rigid experimentalist. He published a number of experiments of his own, some of which were very ingenious, to show that if you went to work in a proper way, there was a truth in the doctrine of spontaneous generation. Well, it was one of the most fortunate things in the world that M. Pouchet took up this question, because it induced a distinguished French chemist, M. Pasteur, to take up the question on the other side; and he has certainly worked it out in the most perfect manner. I am glad to say, too, that he has published his researches in time to enable me to give you an account of them. He verified all the experiments which I have just mentioned to you—and then finding those extraordinary anomalies, as in the case of the mercury bath and the milk, he set himself to work to discover their nature. In the case of milk he found it to be a question of temperature. Milk in a fresh state is slightly alkaline; and it is a very curious circumstance, but this very slight degree of alkalinity seems to have the effect of preserving the organisms which fall into it from the air from being destroyed at a temperature of 212°, which is the boiling point. But if you raise the temperature 10° when you boil it, the milk behaves like everything else; and if the air with which it comes in contact, after being boiled at this temperature, is passed through a red-hot tube, you will not get a trace of organisms.

He then turned his attention to the mercury bath, and found on examination that the surface of the mercury was almost always covered with a very fine dust. He found that even the mercury itself was positively full of organic matters; that from being constantly exposed to the air, it had collected an immense number of these infusorial organisms from the air. Well, under these circumstances he felt that the case was quite clear, and that the mercury was not what it had appeared to M. Schwann to be,—a bar to the admission of these organisms; but that, in reality, it acted as a reservoir from which the infusion was immediately supplied with the large quantity that had so puzzled him.

But not content with explaining the experiments of others, M. Pasteur went to work to satisfy himself completely. He said to himself: "If my view is right, and if, in point of fact, all these appearances of spontaneous generation are altogether due to the falling of minute germs suspended in the atmosphere,—why, I ought not only to be able to show the germs, but I ought to be able to catch and sow them, and produce the resulting organisms." He, accordingly, constructed a very ingenious apparatus to enable him to accomplish the trapping of the *"germ dust"* in the air. He fixed in the window of his room a glass tube, in the centre of which he had placed a ball of gun-cotton, which, as you all know, is ordinary cotton-wool, which, from having been steeped in strong acid, is converted into a substance of great explosive power. It is also soluble in alcohol and ether. One end of the glass tube was, of course, open to the external air; and at the other end of it he placed an aspirator, a contrivance for causing a current of the external air to pass through the tube. He kept this apparatus going for four-and-twenty hours, and then removed the *dusted* gun-cotton, and dissolved it in alcohol and ether. He then allowed this to stand for a few hours, and the result was, that a very fine dust was gradually deposited at the bottom of it. That dust, on being transferred to the stage of a microscope, was found to contain an enormous number of starch grains. You know that the materials of our food and the greater portion of plants are composed of starch, and we are constantly making use of it in a variety of ways, so that there is always a quantity of it suspended in the air. It is these starch grains which form many of those bright specks that we see dancing in a ray of light sometimes. But besides these, M. Pasteur found also an immense number of other organic substances such as spores of fungi, which had been floating about in the air and had got caged in this way.

He went farther, and said to himself, "If these really are the things that give rise to the appearance of spontaneous generation, I ought to be able to take a ball of this *dusted* gun-cotton and put it into one of my vessels, containing that boiled infusion which has been kept away from the air, and in which no infusoria are at present developed, and then, if I am right, the introduction of this gun-cotton will give rise to organisms."

Accordingly, he took one of these vessels of infusion, which had been kept eighteen months, without the least appearance of life in it, and by a most ingenious contrivance, he managed to break it open and introduce such a ball of gun-cotton, without allowing the infusion or the cotton ball to come into contact with any air but that which had been subjected to a red heat, and in twenty-four hours he had the satisfaction of finding all the indications of what had been hitherto called spontaneous generation. He had succeeded in catching the germs and developing organisms in the way he had anticipated.

It now struck him that the truth of his conclusions might be demonstrated without all the apparatus he had employed. To do this, he took some decaying animal or vegetable substance, such as urine, which is an extremely decomposable substance, or the juice of yeast, or perhaps some other artificial preparation, and filled a vessel having a long tubular neck, with it. He then boiled the liquid and bent that long neck into an S shape or zig-zag, leaving it open at the end. The infusion then gave no trace of any appearance of spontaneous generation, however long it might be left, as all the germs in the air were deposited in the beginning of the bent neck. He then cut the tube close to the vessel, and allowed the ordinary air to have free and direct access; and the result of that was the appearance of organisms in it, as soon as the infusion had been allowed to stand long enough to allow of the growth of those it received from the air, which was about forty-eight hours. The result of M. Pasteur's experiments proved, therefore, in the most conclusive manner, that all the appearances of spontaneous generation arose from nothing more than the deposition of the germs of organisms which were constantly floating in the air.

To this conclusion, however, the objection was made, that if that were the cause, then the air would contain such an enormous number of these germs, that it would be a continual fog. But M. Pasteur replied that they are not there in anything like the number we might suppose, and that an exaggerated view has been held on that subject; he showed that the chances of animal or vegetable life appearing in infusions, depend entirely on the conditions under which they are exposed. If they are exposed to the ordinary atmosphere around us, why, of course, you may have organisms appearing early. But, on the other hand, if they are exposed to air at a great height, or in some very quiet cellar, you will often not find a single trace of life.

So that M. Pasteur arrived at last at the clear and definite result, that all these appearances are like the case of the worms in the piece of meat, which was refuted by Redi, simply germs carried by the air and deposited in the liquids in which they afterwards appear. For my own part, I conceive that, with the particulars of M. Pasteur's experiments before us, we cannot fail to arrive at his conclusions; and that the doctrine of spontaneous generation has received a final *coup de grâce*.

You, of course, understand that all this in no way interferes with the *possibility* of the fabrication of organic matters by the direct method to which I have referred, remote as that possibility may be.

LECTURE IV.

THE PERPETUATION OF LIVING BEINGS, HEREDITARY TRANSMISSION AND VARIATION.

The inquiry which we undertook, at our last meeting, into the state of our knowledge of the causes of the phenomena of organic nature,—of the past and of the present,—resolved itself into two subsidiary inquiries: the first was, whether we know anything, either historically or experimentally, of the mode of origin of living beings; the second subsidiary inquiry was, whether, granting the origin, we know anything about the perpetuation and modifications of the forms of organic beings. The reply which I had to give to the first question was altogether negative, and the chief result of my last lecture was, that, neither historically nor experimentally, do we at present know anything whatsoever about the origin of living forms. We saw that, historically, we are not likely to know anything about it, although we may perhaps learn something experimentally; but that at present we are an enormous distance from the goal I indicated.

I now, then, take up the next question, What do we

know of the reproduction, the perpetuation, and the modifications of the forms of living beings, supposing that we have put the question as to their origination on one side, and have assumed that at present the causes of their origination are beyond us, and that we know nothing about them? Upon this question the state of our knowledge is extremely different; it is exceedingly large: and, if not complete, our experience is certainly most extensive. It would be impossible to lay it all before you, and the most I can do, or need do to-night, is to take up the principal points and put them before you with such prominence as may subserve the purposes of our present argument.

The method of the perpetuation of organic beings is of two kinds,—the asexual and the sexual. In the first the perpetuation takes place from and by a particular act of an individual organism, which sometimes may not be classed as belonging to any sex at all. In the second case, it is in consequence of the mutual action and interaction of certain portions of the organisms of usually two distinct individuals,—the male and the female. The cases of asexual perpetuation are by no means so common as the cases of sexual perpetuation; and they are by no means so common in the animal as in the vegetable world. You are all probably familiar with the fact, as a matter of experience, that you can propagate plants by means of what are called "cuttings;" for example, that by taking a cutting from a geranium plant, and rearing it properly, by supplying it with light and warmth and nourishment from the earth, it grows up and takes the form of its parent, having all the properties and peculiarities of the original plant.

Sometimes this process, which the gardener performs artificially, takes place naturally; that is to say, a little bulb, or portion of the plant, detaches itself, drops off, and becomes capable of growing as a separate thing. That is the case with many bulbous plants, which throw off in this way secondary bulbs, which are lodged in the ground and become developed into plants. This is an asexual process, and from it results the repetition or reproduction of the form of the original being from which the bulb proceeds.

Among animals the same thing takes place. Among the lower forms of animal life, the infusorial animalculæ we have already spoken of throw off certain portions, or break themselves up in various directions, sometimes transversely or sometimes longitudinally; or they may give off buds, which detach themselves and develop into their proper forms. There is the common fresh-water Polype, for instance, which multiplies itself in this way. Just in the same way as the gardener is able to multiply and reproduce the peculiarities and characters of particular plants by means of cuttings, so can the physiological experimentalist,—as was shown by the Abbé Trembley many years ago,—so can he do the same thing with many of the lower forms of animal life. M. de Trembley showed that you could take a polype and cut it into two, or four, or many pieces, mutilating it in all directions, and the pieces would still grow up and reproduce completely the original form of the animal. These are all cases of asexual multiplication, and there are other instances, and still more extraordinary ones, in which this process takes place naturally, in a more hidden, a more recondite kind of way. You are all of you familiar with that little green insect, the *Aphis* or blight, as it is called. These little animals, during a very considerable part of their existence, multiply themselves by means of a kind of internal budding, the buds being developed into essentially asexual animals, which are neither male nor female; they become converted into young *Aphides*, which repeat the process, and their offspring after them, and so on again; you may go on for nine or ten, or even twenty or more successions; and there is no very good reason to say how soon it might terminate, or how long it might not go on if the proper conditions of warmth and nourishment were kept up.

Sexual reproduction is quite a distinct matter. Here, in all these cases, what is required is the detachment of two portions of the parental organisms, which portions we know as the egg or the spermatozoon. In plants it is the ovule and the pollen-grain, as in the flowering plants, or the ovule and the antherozooid, as in the flowerless. Among all forms of animal life, the spermatozoa proceed from the male sex, and the egg is the product of the female. Now, what is remarkable about this mode of reproduction is this, that the egg by itself, or the spermatozoa by themselves, are unable to assume the parental form; but if they be brought into contact with one another, the effect of the mixture of organic substances proceeding from two sources appears to confer an altogether new vigour to the mixed product. This process is brought about, as we all know, by the sexual intercourse of the two sexes, and is called the act of impregnation. The result of this act on the part of the male and female is, that the formation of a new being is set up in the ovule or egg; this ovule or egg soon begins to be divided and subdivided, and to be fashioned into various complex organisms, and eventually to develop into the form of one of its parents, as I explained in the first lecture. These are the processes by which the perpetuation of organic beings is secured. Why there should be the two modes—why this re-invigoration should be required on the part of the female element we do not know; but it is most assuredly the fact, and it is presumable, that, however long the process of asexual multiplication could be continued,—I say there is good reason to believe that it would come to an end if a new commencement were not obtained by a conjunction of the two sexual elements.

That character which is common to these two distinct processes is this, that, whether we consider the reproduction, or perpetuation, or modification of organic beings as they take place asexually, or as they may take place sexually,—in either case, I say, the offspring has a constant tendency to assume, speaking generally, the character of the parent. As I said just now, if you take a slip of a plant, and tend it with care, it will eventually grow up and develop into a plant like that from which it had sprung; and this tendency is so strong that, as gardeners know, this mode of multiplying by means of cuttings is the only secure mode of propagating very many varieties of plants; the peculiarity of the primitive stock seems to be better preserved if you propagate it by means of a slip than if you resort to the sexual mode.

Again, in experiments upon the lower animals, such as the polype, to which I have referred, it is most extraordinary that, although cut up into various pieces, each particular piece will grow up into the form of the primitive stock; the head, if separated, will reproduce the body and the tail; and if you cut off the tail, you will find that that will reproduce the body and all the rest of the members, without in any way deviating from the plan of the organism from which these portions have been detached. And so far does this go, that some experimentalists have carefully examined the lower orders of animals,—among them the Abbé Spallanzani, who made a number of experiments upon snails and salamanders,—and have found that they might mutilate them to an incredible extent; that you might cut off the jaw or the greater part of the head, or the leg or the tail, and repeat the experiment several times, perhaps, cutting off the same member again and again; and yet each of those types would be reproduced according to the primitive type : nature making no mistake, never putting on a fresh kind of leg, or head, or tail, but always tending to repeat and to return to the primitive type.

It is the same in sexual reproduction : it is a matter of perfectly common experience, that the tendency on the part of the offspring always is, speaking broadly, to reproduce the form of the parents. The proverb has it that the thistle does not bring forth grapes; so, among ourselves, there is always a likeness, more or less marked and distinct, between children and their parents. That is a matter of familiar and ordinary observation. We notice the same thing occurring in the cases of the domestic animals—dogs, for instance, and their offspring. In all these cases of propagation and perpetuation, there seems to be a tendency in the offspring to take the characters of the parental organisms. To that tendency a special name is given—and as I may very often use it, I will write it up here on this black-board that you may remember it—it is called *Atavism*; it expresses this tendency to revert to the ancestral type, and comes from the Latin word *atavus*, ancestor.

Well, this *Atavism* which I shall speak of, is, as I said before, one of the most marked and striking tendencies of organic beings; but, side by side with this hereditary tendency there is an equally distinct and remarkable tendency to variation. The tendency to reproduce the original stock has, as it were, its limits, and side by side with it there is a tendency to vary in certain directions, as if there were two opposing powers working upon the organic being, one tending to take it in a straight line, and the other tending to make it diverge from that straight line, first to one side and then to the other.

So that you see these two tendencies need not precisely contradict one another, as the ultimate result may not always be very remote from what would have been the case if the line had been quite straight.

This tendency to variation is less marked in that mode of propagation which takes place asexually; it is in that mode that the minor characters of animal and vegetable structures are most completely preserved. Still, it will happen sometimes, that the gardener, when he has planted a cutting of some favourite plant, will find, contrary to his expectation, that the slip grows up a little different from the primitive stock — that it produces flowers of a different colour or make, or some deviation in one way or another. This is what is called the 'sporting' of plants.

In animals the phenomena of asexual propagation are so obscure, that at present we cannot be said to know much about them; but if we turn to that mode of perpetuation which results from the sexual process, then we find variation a perfectly constant occurrence, to a certain extent; and, indeed, I think that a certain amount of variation from the primitive stock is the necessary result of the method of sexual propagation itself; for, inasmuch as the thing propagated proceeds from two organisms of different sexes and different makes and temperaments, and as the offspring is to be either of one sex or the other, it is quite clear that it cannot be an exact diagonal of the two, or it would be of no sex at all; it cannot be an exact intermediate form between that of each of its parents—it must deviate to one side or the other. You do not find that the male follows the precise type of the male parent, nor does the female always inherit the precise characteristics of the mother,—there is always a proportion of the female character in the male offspring, and of the male character in the female offspring. That must be quite plain to all of you who have looked at all attentively on your own children or those of your neighbours; you will have noticed how very often it may happen that the son shall exhibit the maternal type of character, or the daughter possess the characteristics of the father's family. There are all sorts of intermixtures and intermediate conditions between the two, where complexion, or beauty, or fifty other different peculiarities belonging to either side of the house, are reproduced in other members of the same family. Indeed, it is sometimes to be remarked in this kind of variation, that the variety belongs, strictly speaking, to neither of the immediate parents; you will see a child in a family who is not like either its father or its mother; but some old person who knew its grandfather or grandmother, or, it may be, an uncle, or, perhaps, even a more distant relative, will see a great similarity between the child and one of these. In this way it constantly happens that the characteristic of some previous member of the family comes out and is reproduced and recognized in the most unexpected manner.

But apart from that matter of general experience, there are some cases which put that curious mixture in a very clear light. You are aware that the offspring of the Ass and the Horse, or rather of the he-Ass and the Mare, is what is called a Mule; and, on the other hand, the offspring of the Stallion and the she-Ass is what is called a *Hinny*. It is a very rare thing in this country to see a Hinny. I never saw one myself; but they have been very carefully studied. Now, the curious thing is this, that although you have the same elements in the experiment in each case, the offspring is entirely different in character, according as the male influence comes from the Ass or the Horse. Where the Ass is the male, as in the case of the Mule, you find that the head is like that of the Ass, that the ears are long, the tail is tufted at the end, the feet are small, and the voice is an unmistakable bray; these are all points of similarity to the Ass; but, on the other hand, the barrel of the body and the cut of the neck are much more like those of the Mare. Then, if you look at the Hinny,—the result of the union of the Stallion and the she-Ass, then you find it is the Horse that has the predominance; that the head is more like that of the Horse, the ears are shorter, the legs coarser, and the type is altogether altered; while the voice, instead of being a bray, is the ordinary neigh of the Horse. Here, you see, is a most curious thing: you take exactly the same elements, Ass and Horse, but you combine the sexes in a different manner, and the result is modified accordingly. You have in this case, however, a result which is not general and universal—there is usually an important preponderance, but not always on the same side.

Here, then, is one intelligible, and, perhaps, necessary cause of variation: the fact, that there are two sexes sharing in the production of the offspring, and that the share taken by each is different and variable, not only for each combination, but also for different members of the same family.

Secondly, there is a variation, to a certain extent,—though in all probability the influence of this cause has been very much exaggerated—but there is no doubt that variation is produced, to a certain extent, by what are commonly known as external conditions,—such as temperature, food, warmth, and moisture. In the long run, every variation depends, in some sense, upon external conditions, seeing that everything has a cause of its own. I use the term "external conditions" now in the sense in which it is ordinarily employed: certain it is, that external conditions have a definite effect. You may take a plant which has single flowers, and by dealing with the soil, and nourishment, and so on, you may by-and-by convert single flowers into double flowers, and make thorns shoot out into branches. You may thicken or make various modifications in the shape of the fruit. In animals, too, you may produce analogous changes in this way, as in the case of that deep bronze colour which persons rarely lose after having passed any length of time in tropical countries. You may also alter the development of the muscles very much, by dint of training; all the world knows that exercise has a great effect in this way; we always expect to find the arm of a blacksmith hard and wiry, and possessing a large development of the brachial muscles. No doubt, training, which is one of the forms of external conditions, converts what are originally only instructions, teachings, into habits, or, in other words, into organizations, to a great extent; but this second cause of variation cannot be considered to be by any means a large one. The third cause that I have to mention, however, is a very extensive one. It is one that, for want of a better name, has been called "spontaneous variation;" which means that when we do not know anything about the cause of phenomena, we call it spontaneous. In the orderly chain of causes and effects in this world, there are very few things of which it can be said with truth that they are spontaneous. Certainly not in these physical matters,—in these there is nothing of the kind,—everything depends on previous conditions. But when we cannot trace the cause of phenomena, we call them spontaneous.

Of these variations, multitudinous as they are, but little is known with perfect accuracy. I will mention to you some two or three cases, because they are very remarkable in themselves, and also because I shall want to use them afterwards. Réaumur, a famous French naturalist, a great many years ago, in an essay which he wrote upon the art of hatching chickens,—which was indeed a very curious essay,—had occasion to speak of variations and monstrosities. One very remarkable case had come under his notice of a variation in the form of a human member, in the person of a Maltese, of the name of Gratio Kelleia, who was born with six fingers upon each hand, and the like number of toes to each of his feet. That was a case of spontaneous variation. Nobody knows why he was born with that number of fingers and toes, and as we don't know, we call it a case of "spontaneous" variation. There is another remarkable case also. I select these, because they happen to have been observed and noted very carefully at the time. It frequently happens that a variation occurs, but the persons who notice it do not take any care in noting down the particulars, until at length, when inquiries come to be made, the exact circumstances are forgotten; and hence, multitudinous as may be such "spontaneous" variations, it is exceedingly difficult to get at the origin of them.

The second case is one of which you may find the whole details in the " Philosophical Transactions " for the year 1813, in a paper communicated by Colonel Humphrey to the President of the Royal Society,— " On a new Variety in the Breed of Sheep," giving an account of a very remarkable breed of sheep, which at one time was well known in the northern states of America, and which went by the name of the Ancon or the Otter breed of sheep. In the year 1791, there was a farmer of the name of Seth Wright in Massachusetts, who had a flock of sheep, consisting of a ram and, I think, of some twelve or thirteen ewes. Of this flock of ewes, one at the breeding-time bore a lamb which was very singularly formed; it had a very long body, very short legs, and those legs were bowed! I will tell you by-and-by how this singular variation in the breed of sheep came to be noted, and to have the prominence that it now has. For the present, I mention only these two cases; but the extent of variation in the breed of animals is perfectly obvious to any one who has studied natural history with ordinary attention, or to any person who compares animals with others of the same kind. It is strictly true that there are never any two specimens which are exactly alike;

however similar, they will always differ in some certain particular.

Now let us go back to Atavism,—to the hereditary tendency I spoke of. What will come of a variation when you breed from it, when Atavism comes, if I may say so, to intersect variation? The two cases of which I have mentioned the history, give a most excellent illustration of what occurs. Gratio Kelleia, the Maltese, married when he was twenty-two years of age, and, as I suppose there were no six-fingered ladies in Malta, he married an ordinary five-fingered person. The result of that marriage was four children; the first, who was christened Salvator, had six fingers and six toes, like his father; the second was George, who had five fingers and toes, but one of them was deformed, showing a tendency to variation; the third was André; he had five fingers and five toes, quite perfect; the fourth was a girl, Marie; she had five fingers and five toes, but her thumbs were deformed, showing a tendency toward the sixth.

These children grew up, and when they came to adult years, they all married, and of course it happened that they all married five-fingered and five-toed persons. Now let us see what were the results. Salvator had four children; they were two boys, a girl, and another boy: the first two boys and the girl were six-fingered and six-toed like their grandfather; the fourth boy had only five fingers and five toes. George had only four children: there were two girls with six fingers and six toes; there was one girl with six fingers and five toes on the right side, and five fingers and five toes on the left side, so that she was half and half. The last, a boy, had five fingers and five toes. The third, Andrè, you will recollect, was perfectly well-formed, and he had many children whose hands and feet were all regularly developed. Marie, the last, who, of course, married a man who had only five fingers, had four children: the first, a boy, was born with six toes, but the other three were normal.

Now observe what very extraordinary phenomena are presented here. You have an accidental variation arising from what you may call a monstrosity; you have that monstrosity tendency or variation diluted in the first instance by an admixture with a female of normal construction, and you would naturally expect that, in the results of such an union, the monstrosity, if repeated, would be in equal proportion with the normal type; that is to say, that the children would be half and half, some taking the peculiarity of the father, and the others being of the purely normal type of the mother; but you see we have a great preponderance of the abnormal type. Well, this comes to be mixed once more with the pure, the normal type, and the abnormal is again produced in large proportion, notwithstanding the second dilution. Now what would have happened if these abnormal types had intermarried with each other; that is to say, suppose the two boys of Salvator had taken it into their heads to marry their first cousins, the two first girls of George, their uncle? You will remember that these are all of the abnormal type of their grandfather. The result would probably have been, that their offspring would have been in every case a further development of that abnormal type. You see it is only in the fourth, in the person of Marie, that the tendency, when it appears but slightly in the second generation, is washed out in the third, while the progeny of André, who escaped in the first instance, escape altogether.

We have in this case a good example of nature's tendency to the perpetuation of a variation. Here it is certainly a variation which carried with it no use or benefit; and yet you see the tendency to perpetuation may be so strong, that, notwithstanding a great admixture of pure blood, the variety continues itself up to the third generation, which is largely marked with it. In this case, as I have said, there was no means of the second generation intermarrying with any but five-fingered persons, and the question naturally suggests itself, What would have been the result of such marriage? Réaumur narrates this case only as far as the third generation. Certainly it would have been an exceedingly curious thing if we could have traced this matter any further; had the cousins intermarried, a six-fingered variety of the human race might have been set up.

To show you that this supposition is by no means an unreasonable one, let me now point out what took place in the case of Seth Wright's sheep, where it happened to be a matter of moment to him to obtain a breed or raise a flock of sheep like that accidental variety that I have described—and I will tell you why. In that part of Massachusetts where Seth Wright was living, the fields were separated by fences, and the sheep, which were very active and robust, would roam abroad, and without much difficulty jump over these fences into other people's farms. As a matter of course, this exuberant activity on the part of the sheep constantly gave rise to all sorts of quarrels, bickerings, and contentions among the farmers of the neighbourhood; so it occurred to Seth Wright, who was, like his successors, more or less 'cute, that if he could get a stock of sheep like those with the bandy legs, they would not be able to jump over the fences so readily; and he acted upon that idea. He killed his old ram, and as soon as the young one arrived at maturity, he bred altogether from it. The result was even more striking than in the human experiment which I mentioned just now. Colonel Humphreys testifies that it always happened that the offspring were either pure Ancons or pure ordinary sheep; that in no case was there any mixing of the Ancons with the others. In consequence of this, in the course of a very few years, the farmer was able to get a very considerable flock of this variety, and a large number of them were spread throughout Massachusetts. Most unfortunately, however—I suppose it was because they were so common—nobody took enough notice of them to preserve their skeletons; and although Colonel Humphreys states that he sent a skeleton to the president of the Royal Society at the same time that he forwarded his paper, and I am afraid that the variety has entirely disappeared; for a short time after these sheep had become prevalent in that district, the Merino sheep were introduced; and as their wool was much more valuable, and as they were a quiet race of sheep, and showed no tendency to trespass or jump over fences, the Otter breed of sheep, the wool of which was inferior to that of the Merino, was gradually allowed to die out.

You see that these facts illustrate perfectly well what may be done if you take care to breed from stocks that are similar to each other. After having got a variation, if, by crossing a variation with the original stock, you multiply that variation, and then take care to keep that variation distinct from the original stock, and make them breed together,—then you may almost certainly produce a race whose tendency to continue the variation is exceedingly strong.

This is what is called "selection;" and it is by exactly the same process as that by which Seth Wright bred his Ancon sheep, that our breeds of cattle, dogs, and fowls, are obtained. There are some possibilities of exception, but still, speaking broadly, I may say that this is the way in which all our varied races of domestic animals have arisen; and you must understand that it is not one peculiarity or one characteristic alone in which animals may vary. There is not a single peculiarity or characteristic of any kind, bodily or mental, in which offspring may not vary to a certain extent from the parent and other animals.

Among ourselves this is well known. The simplest physical peculiarity is mostly reproduced. I know a case of a woman who has the lobe of one of her ears a little flattened. An ordinary observer might scarcely notice it, and yet every one of her children has an approximation to the same peculiarity to some extent. If you look at the other extreme, too, the gravest diseases, such as gout, scrofula, and consumption, may be handed down with just the same certainty and persistence as we noticed in the perpetuation of the bandy legs of the Ancon sheep.

However, these facts are best illustrated in animals, and the extent of the variation, as is well known, is very remarkable in dogs. For example, there are some dogs very much smaller than others; indeed, the variation is so enormous that probably the smallest dog would be about the size of the head of the largest; there are very great variations in the structural forms not only of the skeleton but also in the shape of the skull, and in the proportions of the face and the disposition of the teeth.

The Pointer, the Retriever, Bulldog, and the Terrier, differ very greatly, and yet there is every reason to believe that every one of these races has arisen from the same source,—that all the most important races have arisen by this selective breeding from accidental variation.

A still more striking case of what may be done by selective breeding, and it is a better case, because there is no chance of that partial infusion of error to which I alluded, has been studied very carefully by Mr. Darwin,—the case of the domestic pigeons. I dare say there may be some among you who may be pigeon *fanciers*, and I wish you to understand that in approaching the subject, I would speak with all humility and hesitation, as I regret to say that I am not a pigeon fancier. I know it is a great art and mystery, and a thing upon which a man must not speak lightly; but I shall endeavour, as far as my understanding goes, to give you a summary of the published and unpublished information which I have gained from Mr. Darwin.

Among the enormous variety,—I believe there are somewhere about a hundred and fifty kinds of pigeons,—there are four kinds which may be selected as representing the extremest divergences of one kind from another. Their names are the Carrier, the Pouter, the Fantail, and the Tumbler. In these large diagrams that I have here they are each represented in their relative sizes to each other. This first one is the Carrier; you will notice this large excrescence on its beak; it has a comparatively small head; there is a bare space round the eyes; it has a long neck, a very long beak, very strong legs, large feet, long wings, and so on. The second one is the Pouter, a very large bird, with very long legs and beak. It is called the Pouter because it is in the habit of causing its gullet to swell up by inflating it with air. I should tell you that all pigeons have a tendency to do this at times, but in the Pouter it is carried to an enormous extent. The birds appear to be quite proud of their power of swelling and puffing themselves out in this way; and I think it is about as droll a sight as you can well see to look at a cage full of these pigeons puffing and blowing themselves out in this ridiculous manner.

This diagram is a representation of the third kind

I mentioned—the Fantail. It is, you see, a small bird, with exceedingly small legs and a very small beak. It is most curiously distinguished by the size and extent of its tail, which, instead of containing twelve feathers, may have many more,—say thirty, or even more—I believe there are some with as many as forty-two. This bird has a curious habit of spreading out the feathers of its tail in such a way that they reach forward, and touch its head; and if this can be accomplished, I believe it is looked upon as a point of great beauty.

But here is the last great variety,—the Tumbler; and of that great variety, one of the principal kinds, and one most prized, is the specimen represented here—the short-faced Tumbler. Its beak, you see, is reduced to a mere nothing. Just compare the beak of this one and that of the first one, the Carrier—I believe the orthodox comparison of the head and beak of a thoroughly well-bred Tumbler is to stick an oat into a cherry, and that will give you the proper relative proportions of the beak and head. The feet and legs are exceedingly small, and the bird appears to be quite a dwarf when placed side by side with this great Carrier.

These are differences enough in regard to their external appearance; but these differences are by no means the whole or even the most important of the differences which obtain between these birds. There is hardly a single point of their structure which has not become more or less altered; and to give you an idea of how extensive these alterations are, I have here some very good skeletons, for which I am indebted to my friend Mr. Tegetmeier, a great authority in these matters; by means of which, if you examine them by-and-by, you will be able to see the enormous difference in their bony structures.

I had the privilege, some time ago, of access to some important MSS. of Mr. Darwin, who, I may tell you, has taken very great pains and spent much valuable time and attention on the investigation of these variations, and getting together all the facts that bear upon them. I obtained from these MSS. the following summary of the differences between the domestic breeds of pigeons; that is to say, a notification of the various points in which their organization differs. In the first place, the back of the skull may differ a good deal, and the development of the bones of the face may vary a great deal; the back varies a good deal; the shape of the lower jaw varies; the tongue varies very greatly, not only in correlation to the length and size of the beak, but it seems also to have a kind of independent variation of its own. Then the amount of naked skin round the eyes, and at the base of the beak, may vary enormously; so may the length of the eyelids, the shape of the nostrils, and the length of the neck. I have already noticed the habit of blowing out the gullet, so remarkable in the Pouter, and comparatively so in the others. There are great differences, too, in the size of the female and the male, the shape of the body, the number and width of the processes of the ribs, the development of the ribs, and the size, shape, and development of the breastbone. We may notice, too,—and I mention the fact because it has been disputed by what is assumed to be high authority,—the variation in the number of the sacral vertebræ. The number of these varies from eleven to fourteen, and that without any diminution in the number of the vertebræ of the back or of the tail. Then the number and position of the tail-feathers may vary enormously, and so may the number of the primary and secondary feathers of the wings. Again, the length of the feet and of the beak,—although they have no relation to each other, yet appear to go together,—that is, you have a long beak wherever you have long feet. There are differences also in the periods of the acquirement of the perfect plumage,—the size and shape of the eggs,—the nature of flight, and the powers of flight,—so-called "*homing*" birds having enormous flying powers;* while, on the other hand, the little Tumbler is so called because of its extraordinary faculty of turning head over heels in the air, instead of pursuing a distinct course. And, lastly, the dispositions and voices of the birds may vary. Thus the case of the pigeons shows you that there is hardly a single particular,—whether of instinct, or habit, or bony structure, or of plumage,—of either the internal economy or the external shape, in which some variation or change may not take place, which, by selective breeding, may become perpetuated, and form the foundation of, and give rise to, a new race.

---

* The "*Carrier*," I learn from Mr. Tegetmeier, does not *carry*; a high-bred bird of this breed being but a poor flier. The birds which fly long distances, and come home,—"homing" birds,—and are consequently used as carriers, are not "carriers" in the fancy sense.

If you carry in your mind's eye these four varieties of pigeons, you will bear with you as good a notion as you can have, perhaps, of the enormous extent to which a deviation from a primitive type may be carried by means of this process of selective breeding.

LECTURE V.

THE CONDITIONS OF EXISTENCE AS AFFECTING THE PERPETUATION OF LIVING BEINGS.

In the last Lecture I endeavoured to prove to you that, while, as a general rule, organic beings tend to reproduce their kind, there is in them, also, a constantly recurring tendency to vary—to vary to a greater or to a less extent. Such a variety, I pointed out to you, might arise from causes which we do not understand; we therefore called it spontaneous; and it might come into existence as a definite and marked thing, without any gradations between itself and the form which preceded it. I further pointed out, that such a variety having once arisen, might be perpetuated to some extent, and indeed to a very marked extent, without any direct interference, or without any exercise of that process which we called selection. And then I stated further, that by such selection, when exercised artificially—if you took care to breed only from those forms which presented the same peculiarities of any variety which had arisen in this manner—the variation might be perpetuated, as far as we can see, indefinitely.

The next question, and it is an important one for us, is this: Is there any limit to the amount of variation from the primitive stock which can be produced by this process of selective breeding? In considering this question, it will be useful to class the characteristics, in respect of which organic beings vary, under two heads: we may consider structural characteristics, and we may consider physiological characteristics.

In the first place, as regards structural characteristics, I endeavoured to show you, by the skeletons which I had upon the table, and by reference to a great many well-ascertained facts, that the different breeds of Pigeons, the Carriers, Pouters, and Tumblers, might vary in any of their internal and important structural characters to a very great degree; not only might there be changes in the proportions of the skull, and the characters of the feet and beaks, and so on; but that there might be an absolute difference in the number of the vertebræ of the back, as in the sacral vertebræ of the Pouter; and so great is the extent of the variation in these and similar characters that I pointed out to you, by reference to the skeletons and the diagrams, that these extreme varieties may absolutely differ more from one another in their structural characters than do what naturalists call distinct Species of pigeons; that is to say, that they differ so much in structure that there is a greater difference between the Pouter and the Tumbler than there is between such wild and distinct forms as the Rock Pigeon or the Ring Pigeon, or the Ring Pigeon and the Stock Dove; and indeed the differences are of greater value than this, for the structural differences between these domesticated pigeons are such as would be admitted by a naturalist, supposing he knew nothing at all about their origin, to entitle them to constitute even distinct genera.

As I have used this term SPECIES, and shall probably use it a good deal, I had better perhaps devote a word or two to explaining what I mean by it.

Animals and plants are divided into groups, which become gradually smaller, beginning with a KINGDOM, which is divided into SUB-KINGDOMS; then come the smaller divisions called PROVINCES; and so on from a PROVINCE to a CLASS, from a CLASS to an ORDER, from ORDERS to FAMILIES, and from these to GENERA, until we come at length to the smallest groups of animals which can be defined one from the other by constant characters, which are not sexual; and these are what naturalists call SPECIES in practice, whatever they may do in theory.

If in a state of nature you find any two groups of living beings, which are separated one from the other by some constantly-recurring characteristic, I don't care how slight and trivial, so long as it is defined and constant, and does not depend on sexual peculiarities, then all naturalists agree in calling them two species; that is what is meant by the use of the word species—that is to say, it is, for the practical naturalist, a mere question of structural differences.*

We have seen now—to repeat this point once more, and it is very essential that we should rightly under-

---

* I lay stress here on the *practical* signification of "Species." Whether a physiological test between species exist or not, it is hardly ever applicable by the practical naturalist.

stand it—we have seen that breeds, known to have been derived from a common stock by selection, may be as different in their structure from the original stock as species may be distinct from each other.

But is the like true of the physiological characteristics of animals? Do the physiological differences of varieties amount in degree to those observed between forms which naturalists call distinct species? This is a most important point for us to consider.

As regards the great majority of physiological characteristics, there is no doubt that they are capable of being developed, increased, and modified by selection.

There is no doubt that breeds may be made as different as species in many physiological characters. I have already pointed out to you very briefly the different habits of the breeds of Pigeons, all of which depend upon their physiological peculiarities,—as the peculiar habit of tumbling, in the Tumbler,—the peculiarities of flight, in the "homing" birds,—the strange habit of spreading out the tail, and walking in a peculiar fashion, in the Fantail,—and, lastly, the habit of blowing out the gullet, so characteristic of the Pouter. These are all due to physiological modifications, and in all these respects these birds differ as much from each other as any two ordinary species do.

So with Dogs in their habits and instincts. It is a physiological peculiarity which leads the Greyhound to chase its prey by sight,—that enables the Beagle to track it by the scent,—that impels the Terrier to its rat-hunting propensity,—and that leads the Retriever to its habit of retrieving. These habits and instincts are all the results of physiological differences and peculiarities, which have been developed from a common stock, at least there is every reason to believe so. But it is a most singular circumstance, that while you may run through almost the whole series of physiological processes, without finding a check to your argument, you come at last to a point where you do find a check, and that is in the reproductive processes. For there is a most singular circumstance in respect to natural species—at least about some of them—and it would be sufficient for the purposes of this argument, if it were true of only one of them, but there is, in fact, a great number of such cases—and that is, that similar as they may appear to be to mere races or breeds, they present a marked peculiarity in the reproductive process. If you breed from the male and female of the same race, you of course have offspring of the like kind, and if you make the offspring breed together, you obtain the same result, and if you breed from these again, you will still have the same kind of offspring; there is no check. But if you take members of two distinct species, however similar they may be to each other, and make them breed together, you will find a check, with some modifications and exceptions, however, which I shall speak of presently. If you cross two such species with each other, then,—although you may get offspring in the case of the first cross, yet, if you attempt to breed from the products of that crossing, which are what are called HYBRIDS — that is, if you couple a male and a female hybrid — then the result is that in ninety-nine cases out of a hundred you will get no offspring at all: there will be no result whatsoever.

The reason of this is quite obvious in some cases; the male hybrids, although possessing all the external appearances and characteristics of perfect animals, are physiologically imperfect and deficient in the structural parts of the reproductive elements necessary to generation. It is said to be invariably the case with the male mule, the cross between the Ass and the Mare; and hence it is, that, although crossing the Horse with the Ass is easy enough, and is constantly done, as far as I am aware, if you take two mules, a male and a female, and endeavour to breed from them, you get no offspring whatever; no generation will take place. This is what is called the sterility of the hybrids between two distinct species.

You see that this is a very extraordinary circumstance; one does not see why it should be. The common teleological explanation is, that it is to prevent the impurity of the blood resulting from the crossing of one species with another, but you see it does not in reality do anything of the kind. There is nothing in this fact that hybrids cannot breed with each other, to establish such a theory; there is nothing to prevent the Horse breeding with the Ass, or the Ass with the Horse. So that this explanation breaks down, as a great many explanations of this kind do, that are only founded on mere assumptions.

Thus you see that there is a great difference between "mongrels," which are crosses between distinct races, and "hybrids," which are crosses between distinct species. The mongrels are, so far as we know, fertile with one another. But between species, in many cases, you cannot succeed in obtaining even the first cross:

at any rate it is quite certain that the hybrids are often absolutely infertile one with another.

Here is a feature, then, great or small as it may be, which distinguishes natural species of animals. Can we find any approximation to this in the different races known to be produced by selective breeding from a common stock? Up to the present time the answer to that question is absolutely a negative one. As far as we know at present, there is nothing approximating to this check. In crossing the breeds between the Fantail and the Pouter, the Carrier and the Tumbler, or any other variety or race you may name—so far as we know at present—there is no difficulty in breeding together the mongrels. Take the Carrier and the Fantail, for instance, and let them represent the Horse and the Ass in the case of distinct species; then you have, as the result of their breeding, the Carrier-Fantail mongrel,—we will say the male and female mongrel,—and, as far as we know, these two when crossed would not be less fertile than the original cross, or than Carrier with Carrier. Here, you see, is a physiological contrast between the races produced by selective modification and natural species. I shall inquire into the value of this fact, and of some modifying circumstances by and by; for the present I merely put it broadly before you.

But while considering this question of the limitations of species, a word must be said about what is called RECURRENCE—the tendency of races which have been developed by selective breeding from varieties to return to their primitive type. This is supposed by many to put an absolute limit to the extent of selective and all other variations. People say, "It is all very well to talk about producing these different races, but you know very well that if you turned all these birds wild, these Pouters, and Carriers, and so on, they would all return to their primitive stock." This is very commonly assumed to be a fact, and it is an argument that is commonly brought forward as conclusive; but if you will take the trouble to inquire into it rather closely, I think you will find that it is not worth very much. The first question of course is, Do they thus return to the primitive stock? And commonly as the thing is assumed and accepted, it is extremely difficult to get anything like good evidence of it. It is constantly said, for example, that if domesticated Horses are turned wild, as they have been in some parts of Asia Minor and South America, that they return at once to the primitive stock from which they were bred. But the first answer that you make to this assumption is, to ask who knows what the primitive stock was; and the second answer is, that in that case the wild Horses of Asia Minor ought to be exactly like the wild Horses of South America. If they are both like the same thing, they ought manifestly to be like each other! The best authorities, however, tell you that it is quite different. The wild Horse of Asia is said to be of a dun colour, with a largish head, and a great many other peculiarities; while the best authorities on the wild Horses of South America tell you that there is no similarity between their wild Horses and those of Asia Minor; the cut of their heads is very different, and they are commonly chestnut or bay-coloured. It is quite clear, therefore, that as by these facts there ought to have been two primitive stocks, they go for nothing in support of the assumption that races recur to one primitive stock, and so far as this evidence is concerned, it falls to the ground.

Suppose for a moment that it were so, and that domesticated races, when turned wild, did return to some common condition, I cannot see that this would prove much more than that similar conditions are likely to produce similar results; and that when you take back domesticated animals into what we call natural conditions, you do exactly the same thing as if you carefully undid all the work you had gone through, for the purpose of bringing the animal from its wild to its domesticated state. I do not see anything very wonderful in the fact, if it took all that trouble to get it from a wild state, that it should go back into its original state as soon as you removed the conditions which produced the variation to the domesticated form. There is an important fact, however, forcibly brought forward by Mr. Darwin, which has been noticed in connection with the breeding of domesticated pigeons; and it is, that however different these breeds of pigeons may be from each other, and we have already noticed the great differences in these breeds, that if, among any of those variations, you chance to have a blue pigeon turn up, it will be sure to have the black bars across the wings, which are characteristic of the original wild stock, the Rock Pigeon.

Now, this is certainly a very remarkable circumstance; but I do not see myself how it tells very strongly either one way or the other. I think, in fact, that this argument in favour of recurrence to the primitive type might prove a great deal too much for those who so constantly bring it forward. For example,

Mr. Darwin has very forcibly urged, that nothing is commoner than if you examine a dun horse—and I had an opportunity of verifying this illustration lately, while in the islands of the West Highlands, where there are a great many dun horses—to find that horse exhibit a long black stripe down his back, very often stripes on his shoulder, and very often stripes on his legs. I, myself, saw a pony of this description a short time ago, in a baker's cart, near Rothesay, in Bute : it had the long stripe down the back, and stripes on the shoulders and legs, just like those of the Ass, the Quagga, and the Zebra. Now, if we interpret the theory of recurrence as applied to this case, might it not be said that here was a case of a variation exhibiting the characters and conditions of an animal occupying something like an intermediate position between the Horse, the Ass, the Quagga, and the Zebra, and from which these had been developed? In the same way with regard even to Man. Every anatomist will tell you that there is nothing commoner, in dissecting the human body, than to meet with what are called muscular variations—that is, if you dissect two bodies very carefully, you will probably find that the modes of attachment and insertion of the muscles are not exactly the same in both, there being great peculiarities in the mode in which the muscles are arranged ; and it is very singular, that in some dissections of the human body you will come upon arrangements of the muscles very similar indeed to the same parts in the Apes. Is the conclusion in that case to be, that this is like the black bars in the case of the Pigeon, and that it indicates a recurrence to the primitive type from which the animals have been probably developed? Truly, I think that the opponents of modification and variation had better leave the argument of recurrence alone, or it may prove altogether too strong for them.

To sum up,—the evidence as far as we have gone is against the argument as to any limit to divergences, so far as structure is concerned; and in favour of a physiological limitation. By selective breeding we can produce structural divergences as great as those of species, but we cannot produce equal physiological divergences. For the present I leave the question there.

Now, the next problem that lies before us—and it is an extremely important one—is this: Does this selective breeding occur in nature? Because, if there is no proof of it, all that I have been telling you goes for nothing in accounting for the origin of species. Are natural causes competent to play the part of selection in perpetuating varieties? Here we labour under very great difficulties. In the last lecture I had occasion to point out to you the extreme difficulty of obtaining evidence even of the first origin of those varieties which we know to have occurred in domesticated animals. I told you, that almost always the origin of these varieties is overlooked, so that I could only produce two or three cases, as that of Gratio Kelleia and of the Ancon sheep. People forget, or do not take notice of them until they come to have a prominence; and if that is true of artificial cases, under our own eyes, and in animals in our own care, how much more difficult it must be to have at first hand good evidence of the origin of varieties in nature! Indeed, I do not know that it is possible by direct evidence to prove the origin of a variety in nature, or to prove selective breeding; but I will tell you what we can prove—and this comes to the same thing—that varieties exist in nature within the limits of species, and, what is more, that when a variety has come into existence in nature, there are natural causes and conditions, which are amply competent to play the part of a selective breeder; and although that is not quite the evidence that one would like to have— though it is not direct testimony—yet it is exceeding good and exceedingly powerful evidence in its way.

As to the first point, of varieties existing among natural species, I might appeal to the universal experience of every naturalist, and of any person who has ever turned any attention at all to the characteristics of plants and animals in a state of nature; but I may as well take a few definite cases, and I will begin with Man himself.

I am one of those who believe that, at present, there is no evidence whatever for saying, that mankind sprang originally from any more than a single pair; I must say, that I cannot see any good ground whatever, or even any tenable sort of evidence, for believing that there is more than one species of Man. Nevertheless, as you know, just as there are numbers of varieties in animals, so there are remarkable varieties of men. I speak not merely of those broad and distinct variations which you see at a glance. Everybody, of course, knows the difference between a Negro and a white man, and can tell a Chinaman from an Englishman. They each have peculiar characteristics of colour and physiognomy; but you must recollect that the characters of these races go very far deeper—they extend to the bony structure, and to the characters of that most important of all organs to us—the brain; so that, among men belonging to different races, or even within the same race, one man shall have a brain a third, or half, or even seventy per cent. bigger than another; and if you take the whole range of human brains, you will find a variation in some cases of a hundred per cent. Apart from these variations in the size of the brain, the characters of the skull vary. Thus if I draw the figures of a Mongul and of a Negro head on the blackboard, in the case of the last the breadth would be about seven-tenths, and in the other it would be nine-tenths of the total length. So that you see there is abundant evidence of variation among men in their natural condition. And if you turn to other animals there is just the same thing. The fox, for example, which has a very large geographical distribution all over Europe, and parts of Asia, and on the American Continent, varies greatly. There are mostly large foxes in the North, and smaller ones in the South. In Germany alone, the foresters reckon some eight different sorts.

Of the tiger, no one supposes that there is more than one species; they extend from the hottest parts of Bengal, into the dry, cold, bitter steppes of Siberia, into a latitude of 50°,—so that they may even prey upon the reindeer. These tigers have exceedingly different characteristics, but still they all keep their general features, so that there is no doubt as to their being tigers. The Siberian tiger has a thick fur, a small mane, and a longitudinal stripe down the back, while the tigers of Java and Sumatra differ in many important respects from the tigers of Northern Asia. So lions vary; so birds vary; and so, if you go further back and lower down in creation, you find that fishes vary. In different streams, in the same country even, you will find the trout to be quite different to each other and easily recognizable by those who fish in the particular streams. There is the same differences in leeches; leech collectors can easily point out to you the differences and the peculiarities which you yourself would probably pass by; so with fresh-water mussels; so, in fact, with every animal you can mention.

In plants there is the same kind of variation. Take such a case even as the common bramble. The botanists are all at war about it; some of them wanting to make out that there are many species of it, and others maintaining that they are but many varieties of one species; and they cannot settle to this day which is a species and which is a variety!

So that there can be no doubt whatsoever that any plant and any animal may vary in nature; that varieties may arise in the way I have described,—as spontaneous varieties,—and that those varieties may be perpetuated in the same way that I have shown you spontaneous varieties are perpetuated; I say, therefore, that there can be no doubt as to the origin and perpetuation of varieties in nature.

But the question now is:—Does selection take place in nature? is there anything like the operation of man in exercising selective breeding, taking place in nature? You will observe that, at present, I say nothing about species; I wish to confine myself to the consideration of the production of those natural races which everybody admits to exist. The question is, whether in nature there are causes competent to produce races, just in the same way as man is able to produce, by selection, such races of animals as we have already noticed.

When a variety has arisen, the CONDITIONS OF EXISTENCE are such as to exercise an influence which is exactly comparable to that of artificial selection. By Conditions of Existence I mean two things, — there are conditions which are furnished by the physical, the inorganic world, and there are conditions of existence which are furnished by the organic world. There is, in the first place, CLIMATE; under that head I include only temperature and the varied amount of moisture of particular places. In the next place there is what is technically called STATION, which means — given the climate, the particular kind of place in which an animal or a plant lives or grows; for example, the station of a fish is in the water, of a fresh-water fish in fresh water; the station of a marine fish is in the sea, and a marine animal may have a station higher or deeper. So again with land animals: the differences in their stations are those of different soils and neighbourhoods; some being best adapted to a calcareous, and others to an arenaceous soil. The third condition of existence is FOOD, by which I mean food in the broadest sense, the supply of the materials necessary to the existence of an organic being; in the case of a plant the inorganic matters, such as carbonic acid, water, ammonia, and the earthy salts or salines; in the case of the animal the inorganic and organic matters, which we have seen they require; then these are all, at least the two first, what we may call the inorganic or physical conditions of existence. Food takes a midplace, and then come the organic conditions; by which I mean the conditions which depend upon the state of the rest of the organic creation, upon the number and kind of living beings, with which an animal is surrounded. You may class these under two heads: there are organic beings, which operate as *opponents*, and there are organic beings which operate as *helpers* to any given organic creature. The opponents may be of two kinds: there are the *indirect opponents*, which are what we may call *rivals;* and there are the *direct opponents*, those which strive to destroy the creature; and these we call *enemies*. By rivals I mean, of course, in the case of plants, those which require for their support the same kind of soil and station, and, among animals, those which require the same kind of station, or food, or climate; those are the indirect opponents; the direct opponents are, of course, those which prey upon an animal or vegetable. The *helpers* may also be regarded as direct and indirect: in the case of a carnivorous animal, for example, a particular herbaceous plant may in multiplying be an indirect helper, by enabling the herbivora on which the carnivore preys to get more food, and thus to nourish the carnivore more abundantly; the direct helper may be best illustrated by reference to some parasitic creature, such as the tapeworm. The tape-worm exists in the human intestines, so that the fewer there are of men the fewer there will be of tape-worms, other things being alike. It is a humiliating reflection, perhaps, that we may be classed as direct helpers to the tape-worm, but the fact is so: we can all see that if there were no men there would be no tape-worms.

It is extremely difficult to estimate, in a proper way, the importance and the working of the Conditions of Existence. I do not think there were any of us who had the remotest notion of properly estimating them until the publication of Mr. Darwin's work, which has placed them before us with remarkable clearness; and I must endeavour, as far as I can in my own fashion, to give you some notion of how they work. We shall find it easiest to take a simple case, and one as free as possible from every kind of complication.

I will suppose, therefore, that all the habitable part of this globe — the dry land, amounting to about 51,000,000 square miles, — I will suppose that the whole of that dry land has the same climate, and that it is composed of the same kind of rock or soil, so that there will be the same station everywhere; we thus get rid of the peculiar influence of different climates and stations. I will then imagine that there shall be but one organic being in the world, and that shall be a plant. In this we start fair. Its food is to be carbonic acid, water and ammonia, and the saline matters in the soil, which are, by the supposition, everywhere alike. We take one single plant, with no opponents, no helpers, and no rivals; it is to be a "fair field, and no favour." Now, I will ask you to imagine further that it shall be a plant which shall produce every year fifty seeds, which is a very moderate number for a plant to produce; and that, by the action of the winds and currents, these seeds shall be equally

PERPETUATION OF LIVING BEINGS. 123

and gradually distributed over the whole surface of the land. I want you now to trace out what will occur, and you will observe that I am not talking fallaciously any more than a mathematician does when he expounds his problem. If you show that the conditions of your problem are such as may actually occur in nature and do not transgress any of the known laws of nature in working out your proposition, then you are as safe in the conclusion you arrive at as is the mathematician in arriving at the solution of his problem. In science, the only way of getting rid of the complications with which a subject of this kind is environed, is to work in this deductive method. What will be the result, then? I will suppose that every plant requires one square foot of ground to live upon; and the result will be that, in the course of nine years, the plant will have occupied every single available spot in the whole globe! I have chalked upon the blackboard the figures by which I arrive at the result:—

		Plants.
1 × 50 in 1st year =		50
50 × 50 „ 2nd „ =		2,500
2,500 × 50 „ 3rd „ =		125,000
125,000 × 50 „ 4th „ =		6,250,000
6,250,000 × 50 „ 5th „ =		312,500,000
312,500,000 × 50 „ 6th „ =		15,625,000,000
15,625,000,000 × 50 „ 7th „ =		781,250,000,000
781,250,000,000 × 50 „ 8th „ =		39,062,500,000,000
39,062,500,000,000 × 50 „ 9th „ =		1,953,125,000,000,000

51,000,000 sq. miles—the dry surface of the earth × 27,878,400— the number of sq. ft. in 1 sq. mile } = sq.ft. 1,421,798,400,000,000

being 531,326,600,000,000 square feet less than would be required at the end of the ninth year.

You will see from this that, at the end of the first year the single plant will have produced fifty more of its kind; by the end of the second year these will have increased to 2,500; and so on, in succeeding years, you get beyond even trillions; and I am not at all sure that I could tell you what the proper arithmetical denomination of the total number really is; but, at any rate, you will understand the meaning of all those noughts. Then you see that, at the bottom, I have taken the 51,000,000 of square miles, constituting the surface of the dry land; and as the number of square feet are placed under and subtracted from the number of seeds that would be produced in the ninth year, you can see at once that there would be an immense number more of plants than there would be square feet of ground for their accommodation. This is certainly quite enough to prove my point; that between the eighth and ninth year after being planted the single plant would have stocked the whole available surface of the earth.

This is a thing which is hardly conceivable—it seems hardly imaginable—yet it is so. It is indeed simply the law of Malthus exemplified. Mr. Malthus was a clergyman, who worked out this subject most minutely and truthfully some years ago; he showed quite clearly,—and although he was much abused for his conclusions at the time, they have never yet been disproved and never will be—he showed that in consequence of the increase in the number of organic beings in a geometrical ratio, while the means of existence cannot be made to increase in the same ratio, that there must come a time when the number of organic beings will be in excess of the power of production of nutriment, and that thus some check must arise to the further increase of those organic beings. At the end of the ninth year we have seen that each plant would not be able to get its full square foot of ground, and at the end of another year it would have to share that space with fifty others the produce of the seeds which it would give off.

What, then, takes place? Every plant grows up, flourishes, occupies its square foot of ground, and gives off its fifty seeds; but notice this, that out of this number only one can come to anything; there is thus, as it were, forty-nine chances to one against its growing up; it depends upon the most fortuitous circumstances whether any one of these fifty seeds shall grow up and flourish, or whether it shall die and perish. This is what Mr. Darwin has drawn attention to, and called the "STRUGGLE FOR EXISTENCE;" and I have taken this simple case of a plant because some people imagine that the phrase seems to imply a sort of fight.

I have taken this plant and shown you that this is the result of the ratio of the increase, the necessary result of the arrival of a time coming for every species when exactly as many members must be destroyed as are born; that is the inevitable ultimate result of the rate of production. Now, what is the result of all this? I have said that there are forty-nine struggling against every one; and it amounts to this, that the smallest possible start given to any one seed may give it an advantage which will enable it to get ahead of all the others; anything that will enable any one of these seeds to germinate six hours before any of the others will, other things being alike, enable it to choke them out altogether. I have shown you that there is no particular in which plants will not vary from each other; it is quite possible that one of our imaginary plants may vary in such a character as the thickness of the integument of its seeds; it might happen that one of the plants might produce seeds having a thinner integument, and that would enable the seeds of that plant to germinate a little quicker than those of any of the others, and those seeds would most inevitably extinguish the forty-nine times as many that were struggling with them.

I have put it in this way, but you see the practical result of the process is the same as if some person had nurtured the one and destroyed the other seeds. It does not matter how the variation is produced, so long as it is once allowed to occur. The variation in the plant once fairly started tends to become hereditary and reproduce itself; the seeds would spread themselves in the same way and take part in the struggle with the forty-nine hundred, or forty-nine thousand, with which they might be exposed. Thus, by degrees, this variety with some slight organic change or modification, must spread itself over the whole surface of the habitable globe, and extirpate or replace the other kinds. That is what is meant by Natural Selection; that is the kind of argument by which it is perfectly demonstrable that the conditions of existence may play exactly the same part for natural varieties as man does for domesticated varieties. No one doubts at all that particular circumstances may be more favourable for one plant and less so for another, and the moment you admit that, you admit the selective power of nature. Now, although I have been putting a hypothetical case, you must not suppose that I have been reasoning hypothetically. There are plenty of direct experiments which bear out what we may call the theory of natural selection; there is extremely good authority for the statement that if you take the seed of mixed varieties of wheat and sow it, collecting the seed next year and sowing it again, at length you will find that out of all your varieties only two or three have lived, or perhaps even only one. There were one or two varieties which were best fitted to get on, and they have killed out the other kinds in just the same way and with just the same certainty as if you had taken the trouble to remove them. As I have already said, the operation of nature is exactly the same as the artificial operation of man.

But if this be true of that simple case, which I put before you, where there is nothing but the rivalry of one member of a species with others, what must be the operation of selective conditions, when you recollect as a matter of fact, that for every species of animal or plant there are fifty or a hundred species which might all, more or less, be comprehended in the same climate, food, and station;—that every plant has multitudinous animals which prey upon it, and which are its direct opponents; and that these have other animals preying upon them, —that every plant has its indirect helpers in the birds that scatter abroad its seed, and the animals that manure it with their dung;—I say, when these things are considered, it seems impossible that any variation which may arise in a species in nature should not tend in some way or other either to be a little better or worse than the previous stock; if it is a little better it will have an advantage over and tend to extirpate the latter in this crush and struggle; and if it is a little worse it will itself be extirpated.

I know nothing that more appropriately expresses this, than the phrase, "the struggle for existence;" because it brings before your minds, in a vivid sort of way, some of the simplest possible circumstances connected with it. When a struggle is intense there must be some who are sure to be trodden down, crushed, and overpowered by others; and there will be some who just manage to get through only by the help of the slightest accident. I recollect reading an account of the famous retreat of the French troops, under Napoleon, from Moscow. Worn out, tired, and dejected, they at length came to a great river over which there was but one bridge for the passage of the vast army. Disorganized and demoralized as that army was, the struggle must certainly have been a terrible one—every one heeding only himself, and crushing through the ranks and treading down his fellows. The writer of the narrative, who was himself one of those who were fortunate enough to succeed in getting over, and not among the thousands who were left behind or forced into the river, ascribed his escape to the fact that he saw striding onward through the mass a great strong fellow, —one of the French Cuirassiers, who had on a large blue cloak—and he had enough presence of mind to catch and retain a hold of this strong man's cloak. He says, "I caught hold of his cloak, and although he swore at me and cut at and struck me by turns, and at last, when he found he could not shake me off, fell to entreating me to leave go or I should prevent him from escaping, besides not assisting myself, I still kept tight hold of him, and would not quit my grasp until he had at last dragged me through." Here you see was a case of selective saving—if we may so term it—depending for its success on the strength of the cloth of the Cuirassier's cloak. It is the same in nature; every species has its bridge of Beresina; it has to fight its way through and struggle with other species; and when well nigh overpowered, it may be that the smallest chance, something in its colour, perhaps—the minutest circumstance—will turn the scale one way or the other.

Suppose that by a variation of the black race it had produced the white man at any time—you know that the Negroes are said to believe this to have been the case, and to imagine that Cain was the first white man, and that we are his descendants — suppose that this had ever happened, and that the first residence of this human being was on the West Coast of Africa. There is no great structural difference between the white man and the Negro, and yet there is something so singularly different in the constitution of the two, that the malarias of that country, which do not hurt the black at all, cut off and destroy the white. Then you see there would have been a selective operation performed; if the white man had risen in that way, he would have been selected out and removed by means of the malaria. Now there really is a very curious case of selection of this sort among pigs, and it is a case of selection of colour, too.

I

In the woods of Florida there are a great many pigs, and it is a very curious thing that they are all black, every one of them. Professor Wyman was there some years ago, and on noticing no pigs but these black ones, he asked some of the people how it was that they had no white pigs, and the reply was that in the woods of Florida there was a root which they called the Paint Root, and that if the white pigs were to eat any of it, it had the effect of making their hoofs crack, and they died, but if the black pigs eat any of it, it did not hurt them at all. Here was a very simple case of natural selection. A skilful breeder could not more carefully develope the black breed of pigs, and weed out all the white pigs, than the Paint Root does.

To show you how remarkably indirect may be such natural selective agencies as I have referred to, I will conclude by noticing a case mentioned by Mr. Darwin, and which is certainly one of the most curious of its kind. It is that of the Humble Bee. It has been noticed that there are a great many more humble bees in the neighbourhood of towns, than out in the open country; and the explanation of the matter is this: the humble bees build nests, in which they store their honey and deposit the larvæ and eggs. The field mice are amazingly fond of the honey and larvæ; therefore, wherever there are plenty of field mice, as in the country, the humble bees are kept down; but in the neighbourhood of towns, the number of cats which prowl about the fields eat up the field mice, and of course the more mice they eat up the less there are to prey upon the larvæ of the bees—the cats are therefore the INDIRECT

HELPERS of the bees.* Coming back a step farther we may say that the old maids are also indirect friends of the humble bees, and indirect enemies of the field mice, as they keep the cats which eat up the latter! This is an illustration somewhat beneath the dignity of the subject, perhaps, but it occurs to me in passing, and with it I will conclude this lecture.

* The humble bees, on the other hand, are direct helpers of some plants, such as the heartsease and red clover, which are fertilized by the visits of the bees; and they are indirect helpers of the numerous insects which are more or less completely supported by the heartsease and red clover.

LECTURE VI.

A CRITICAL EXAMINATION OF THE POSITION OF MR. DARWIN'S WORK, "ON THE ORIGIN OF SPECIES," IN RELATION TO THE COMPLETE THEORY OF THE CAUSES OF THE PHENOMENA OF ORGANIC NATURE.

In the preceding five lectures I have endeavoured to give you an account of those facts, and of those reasonings from facts, which form the data upon which all theories regarding the causes of the phenomena of organic nature must be based. And, although I have had frequent occasion to quote Mr. Darwin—as all persons hereafter, in speaking upon these subjects, will have occasion to quote his famous book on the "Origin of Species,"—you must yet remember that, wherever I have quoted him, it has not been upon theoretical points, or for statements in any way connected with his particular speculations, but on matters of fact, brought forward by himself, or collected by himself, and which appear incidentally in his book. If a man *will* make a book, professing to discuss a single question, an encyclopædia, I cannot help it.

Now, having had an opportunity of considering in this sort of way the different statements bearing upon all theories whatsoever, I have to lay before you, as fairly as I can, what is Mr. Darwin's view of the matter and what position his theories hold, when judged by the principles which I have previously laid down, as deciding our judgments upon all theories and hypotheses.

I have already stated to you that the inquiry respecting the causes of the phenomena of organic nature resolves itself into two problems—the first being the question of the origination of living or organic beings; and the second being the totally distinct problem of the modification and perpetuation of organic beings when they have already come into existence. The first question Mr. Darwin does not touch; he does not deal with it at all; but he says :—" Given the origin of organic matter —supposing its creation to have already taken place, my object is to show in consequence of what laws and what demonstrable properties of organic matter, and of its environments, such states of organic nature as those with which we are acquainted must have come about." This, you will observe, is a perfectly legitimate proposition; every person has a right to define the limits of the inquiry which he sets before himself; and yet it is a most singular thing that in all the multifarious, and, not unfrequently, ignorant attacks which have been made upon the "Origin of Species," there is nothing which has been more speciously criticised than this particular limitation. If people have nothing else to urge against the book, they say—" Well, after all, you see Mr. Darwin's explanation of the 'Origin of Species' is not good for much, because, in the long run, he admits that he does not know how organic matter began to exist. But if you admit any special creation for the first particle of organic matter you may just as well admit it for all the rest; five hundred or five thousand distinct creations are just as intelligible, and just as little difficult to understand, as one." The answer to these cavils is two-fold. In the first place, all human inquiry must stop somewhere; all our knowledge and all our investigation cannot take us beyond the limits set by the finite and restricted character of our faculties, or destroy the endless unknown, which accompanies, like its shadow, the endless procession of phenomena. So far as I can venture to offer an opinion on such a matter, the purpose of our being in existence, the highest object that human beings can set before themselves, is not the pursuit of any such chimera as the annihilation of the unknown; but it is simply the unwearied endeavour to remove its boundaries a little further from our little sphere of action.

I wonder if any historian would for a moment admit the objection, that it is preposterous to trouble ourselves about the history of the Roman Empire, because we do not know anything positive about the origin and first building of the city of Rome! Would it be a fair objection to urge, respecting the sublime discoveries of a Newton, or a Kepler, those great philosophers, whose discoveries have been of the profoundest benefit and service to all men,—to say to them—"After all that you have told us as to how the planets revolve, and how they are maintained in their orbits, you cannot tell us what is the cause of the origin of the sun, moon, and stars. So what is the use of what you have done?" Yet these objections would not be one whit more preposterous than the objections which have been made to the "Origin of Species." Mr. Darwin, then, had a perfect right to limit his inquiry as he pleased, and the only question for us—the inquiry being so limited—is to ascertain whether the method of his inquiry is sound or unsound; whether he has obeyed the canons which must guide and govern all investigation, or whether he has broken them; and it was because our inquiry this evening is essentially limited to that question, that I spent a good deal of time in a former lecture (which, perhaps some of you thought might have been better employed) in endeavouring to illustrate the method and nature of scientific inquiry in general. We shall now have to put in practice the principles that I then laid down.

I stated to you in substance, if not in words, that wherever there are complex masses of phenomena to be inquired into, whether they be phenomena of the affairs of daily life, or whether they belong to the more abstruse and difficult problems laid before the philosopher, our course of proceeding in unravelling that complex chain of phenomena with a view to get at its cause, is always the same; in all cases we must invent an hypothesis; we must place before ourselves some more or less likely supposition respecting that cause; and then, having assumed an hypothesis, having supposed a cause for the phenomena in question, we must endeavour, on the one hand, to demonstrate our hypothesis, or, on the other, to upset and reject it altogether, by testing it in three ways. We must, in the first place, be prepared to prove that the supposed causes of the phenomena exist in nature; that they are what the logicians call *vera causæ*—true causes;—in the next place, we should be prepared to show that the assumed causes of the phenomena are competent to produce such phenomena as those which we wish to explain by them; and in the last place, we ought to be able to show that no other known causes are competent to produce these phenomena. If we can succeed in satisfying these three conditions we shall have demonstrated our hypothesis; or rather I ought to say, we shall have proved it as far as certainty is possible for us; for, after all, there is no one of our surest convictions which may not be upset, or at any rate modified by a further accession of knowledge. It was because it satisfied these conditions that we accepted the hypothesis as to the disappearance of the tea-pot and spoons in the case I supposed in a previous lecture; we found that our hypothesis on that subject was tenable and valid, because the supposed cause existed in nature, because it was competent to account for the phenomena, and because no other known cause was competent to account for them; and it is upon similar grounds that any hypothesis you choose to name is accepted in science as tenable and valid.

What is Mr. Darwin's hypothesis? As I apprehend it—for I have put it into a shape more convenient for common purposes than I could find *verbatim* in his book—as I apprehend it, I say, it is, that all the phenomena of organic nature, past and present, result from, or are caused by, the inter-action of those properties of organic matter, which we have called ATAVISM and VARIABILITY, with the CONDITIONS OF EXISTENCE; or, in other words,—given the existence of organic matter, its tendency to transmit its properties, and its tendency occasionally to vary; and, lastly, given the conditions of existence by which organic matter is surrounded—that these put together are the causes of the Present and of the Past conditions of ORGANIC NATURE.

Such is the hypothesis as I understand it. Now let us see how it will stand the various tests which I laid down just now. In the first place, do these supposed causes of the phenomena exist in nature? Is it the fact that in nature these properties of organic matter —atavism and variability—and those phenomena which we have called the conditions of existence,—is it true that they exist? Well, of course, if they do not exist, all that I have told you in the last three or four lectures must be incorrect, because I have been attempting to prove that they do exist, and I take it that there is abundant evidence that they do exist; so far, therefore, the hypothesis does not break down.

But in the next place comes a much more difficult inquiry:—Are the causes indicated competent to give rise to the phenomena of organic nature? I suspect that this is indubitable to a certain extent. It is demonstrable, I think, as I have endeavoured to show you, that they are perfectly competent to give rise to all the phenomena which are exhibited by RACES in nature. Furthermore, I believe that they are quite competent to account for all that we may call purely structural phenomena which are exhibited by SPECIES in nature. On that point also I have already enlarged somewhat. Again, I think that the causes assumed are competent to account for most of the physiological characteristics of species, and I not only think that they are competent to account for them, but I think that they account for many things which otherwise remain wholly unaccountable and inexplicable, and I may say incomprehensible. For a full exposition of the grounds on which this conviction is based, I must refer you to Mr. Darwin's work; all that I can do now is to illustrate what I have said by two or three cases taken almost at random.

I drew your attention, on a previous evening, to the facts which are embodied in our systems of Classification, which are the results of the examination and comparison of the different members of the animal kingdom one with another. I mentioned that the whole of the animal kingdom is divisible into five sub-kingdoms; that each of these sub-kingdoms is again divisible into provinces; that each province may be divided into classes, and the classes into the successively smaller groups, orders, families, genera, and species.

Now, in each of these groups, the resemblance in structure among the members of the group is closer in proportion as the group is smaller. Thus, a man and a worm are members of the animal kingdom in virtue of certain apparently slight though really fundamental resemblances which they present. But a man and a fish are members of the same Sub-kingdom *Vertebrata*, because they are much more like one another than either of them is to a worm, or a snail, or any member of the other sub-kingdoms. For similar reasons men and horses are arranged as members of the same Class, *Mammalia*; men and apes as members of the same

Order, *Primates;* and if there were any animals more like men than they were like any of the apes, and yet different from men in important and constant particulars of their organization, we should rank them as members of the same Family, or of the same Genus, but as of distinct Species.

That it is possible to arrange all the varied forms of animals into groups, having this sort of singular subordination one to the other, is a very remarkable circumstance; but, as Mr. Darwin remarks, this is a result which is quite to be expected, if the principles which he lays down be correct. Take the case of the races which are known to be produced by the operation of atavism and variability, and the conditions of existence which check and modify these tendencies. Take the case of the pigeons that I brought before you: there it was shown that they might be all classed as belonging to some one of five principal divisions, and that within these divisions other subordinate groups might be formed. The members of these groups are related to one another in just the same way as the genera of a family, and the groups themselves as the families of an order, or the orders of a class; while all have the same sort of structural relations with the wild Rock-pigeon, as the members of any great natural group have with a real or imaginary typical form. Now, we know that all varieties of pigeons of every kind have arisen by a process of selective breeding from a common stock, the Rock-pigeon; hence, you see, that if all species of animals have proceeded from some common stock, the general character of their structural relations, and of our systems of classification, which express those relations, would be just what we find them to be. In other words, the hypothetical cause is, so far, competent to produce effects similar to those of the real cause.

Take, again, another set of very remarkable facts,—the existence of what are called rudimentary organs, organs for which we can find no obvious use, in the particular animal economy in which they are found, and yet which are there.

Such are the splint-like bones in the leg of the horse, which I here show you, and which correspond with bones which belong to certain toes and fingers in the human hand and foot. In the horse you see they are quite rudimentary, and bear neither toes nor fingers; so that the horse has only one "finger" in his fore-foot and one "toe" in his hind-foot. But it is a very curious thing that the animals closely allied to the horse show more toes than he; as the rhinoceros, for instance: he has these extra toes well formed, and anatomical facts show very clearly that he is very closely related to the horse indeed. So we may say that animals, in an anatomical sense nearly related to the horse, have those parts which are rudimentary in him, fully developed.

Again, the sheep and the cow have no cutting-teeth, but only a hard pad in the upper jaw. That is the common characteristic of ruminants in general. But the calf has in its upper jaw some rudiments of teeth which never are developed, and never play the part of teeth at all. Well, if you go back in time, you find some of the older, now extinct, allies of the ruminants have well-developed teeth in their upper jaws; and at the present day the pig (which is in structure closely connected with ruminants) has well-developed teeth in its upper jaw; so that here is another instance of organs well developed and very useful, in one animal, represented by rudimentary organs, for which we can discover no purpose whatsoever, in another closely allied animal. The whalebone whale, again, has horny " whalebone " plates in its mouth, and no teeth; but the young fœtal whale, before it is born, has teeth in its jaws; they, however, are never used, and they never come to anything. But other members of the group to which the whale belongs have well-developed teeth in both jaws.

Upon any hypothesis of special creation, facts of this kind appear to me to be entirely unaccountable and inexplicable, but they cease to be so if you accept Mr. Darwin's hypothesis, and see reason for believing that the whalebone whale and the whale with teeth in its mouth both sprang from a whale that had teeth, and that the teeth of the fœtal whale are merely remnants—recollections, if we may so say—of the extinct whale. So in the case of the horse and the rhinoceros: suppose that both have descended by modification from some earlier form which had the normal number of toes, and the persistence of the rudimentary bones which no longer support toes in the horse becomes comprehensible.

In the language that we speak in England, and in the language of the Greeks, there are identical verbal roots, or elements entering into the composition of words. That fact remains unintelligible so long as we suppose English and Greek to be independently created tongues; but when it is shown that both languages are descended from one original, the Sanscrit, we give an explanation of that resemblance. In the same way the existence of identical structural roots, if I may so term them, entering into the composition of widely different animals, is striking evidence in favour of the descent of those animals from a common original.

To turn to another kind of illustration:—If you regard the whole series of stratified rocks—that enormous thickness of sixty or seventy thousand feet that I have mentioned before, constituting the only record we have of a most prodigious lapse of time, that time being, in all probability, but a fraction of that of which we have no record;—if you observe in these successive strata of rocks successive groups of animals arising and dying out, a constant succession, giving you the same kind of impression, as you travel from one group of strata to another, as you would have in travelling from one country to another;—when you find this constant succession of forms, their traces obliterated except to the man of science,—when you look at this wonderful history, and ask what it means, it is only a paltering with words if you are offered the reply,— "They were so created."

But if, on the other hand, you look on all forms of organized beings as the results of the gradual modification of a primitive type, the facts receive a meaning, and you see that these older conditions are the necessary predecessors of the present. Viewed in this light the facts of palæontology receive a meaning—upon any other hypothesis, I am unable to see, in the slightest degree, what knowledge or signification we are to draw out of them. Again, note as bearing upon the same point, the singular likeness which obtains between the successive Faunæ and Floræ, whose remains are preserved on the rocks: you never find any great and enormous difference between the immediately successive Faunæ and Floræ, unless you have reason to believe there has also been a great lapse of time or a great change of conditions. The animals, for instance, of the newest tertiary rocks, in any part of the world, are always, and without exception, found to be closely allied with those which now live in that part of the world. For example, in Europe, Asia, and Africa, the large mammals are at present rhinoceri, hippopotami, elephants, lions, tigers, oxen, horses, &c.; and if you examine the newest tertiary deposits, which contain the animals and plants which immediately preceded those which now exist in the same country, you do not find gigantic specimens of ant-eaters and kangaroos, but you find rhinoceroses, elephants, lions, tigers, &c.,—of different species to those now living,—but still their close allies. If you turn to South America, where, at the present day, we have great sloths and armadilloes and creatures of that kind, what do you find in the newest tertiaries? You find the great sloth-like creature, the *Megatherium*, and the great armadillo, the *Glyptodon*, and so on. And if you go to Australia you find the same law holds good, namely, that that condition of organic nature which has preceded the one which now exists, presents differences perhaps of species, and of genera, but that the great types of organic structure are the same as those which now flourish.

What meaning has this fact upon any other hypothesis or supposition than one of successive modification? But if the population of the world, in any age, is the result of the gradual modification of the forms which peopled it in the preceding age,—if that has been the case, it is intelligible enough; because we may expect that the creature that results from the modification of an elephantine mammal shall be something like an elephant, and the creature which is produced by the modification of an armadillo-like mammal shall be like an armadillo. Upon that supposition, I say, the facts are intelligible; upon any other, that I am aware of, they are not.

So far, the facts of palæontology are consistent with almost any form of the doctrine of progressive modification; they would not be absolutely inconsistent with the wild speculations of De Maillet, or with the less objectionable hypothesis of Lamarck. But Mr. Darwin's views have one peculiar merit; and that is, that they are perfectly consistent with an array of facts which are utterly inconsistent with and fatal to, any other hypothesis of progressive modification which has yet been advanced. It is one remarkable peculiarity of Mr. Darwin's hypothesis that it involves no necessary progression or incessant modification, and that it is perfectly consistent with the persistence for any length of time of a given primitive stock, contemporaneously with its modifications. To return to the case of the domestic breeds of pigeons, for example; you have the Dove-cot pigeon, which closely resembles the Rock pigeon, from which they all started, existing at the same time with the others. And if species are developed in the same way in nature, a primitive stock and its modifications may, occasionally, all find the conditions fitted for their existence; and though they come into competition, to a certain extent, with one another, the derivative species may not necessarily extirpate the primitive one, or *vice versâ*.

Now palæontology shows us many facts which are perfectly harmonious with these observed effects of the process by which Mr. Darwin supposes species to have originated, but which appear to me to be totally inconsistent with any other hypothesis which has been proposed. There are some groups of animals and plants, in the fossil world, which have been said to belong to " persistent types," because they have persisted, with very little change indeed, through a very great range of time, while everything about them has changed largely. There are families of fishes whose type of construction has persisted all the way from the carboniferous rock right up to the cretaceous; and others which have lasted through almost the whole range of the secondary rocks, and from the lias to the older tertiaries. It is something stupendous this —to consider a genus lasting without essential modifications through all this enormous lapse of time while almost everything else was changed and modified.

Thus I have no doubt that Mr. Darwin's hypothesis will be found competent to explain the majority of the phenomena exhibited by species in nature; but in an earlier lecture I spoke cautiously with respect to its power of explaining all the physiological peculiarities of species.

There is, in fact, one set of these peculiarities which the theory of selective modification, as it stands at present, is not wholly competent to explain, and that is the group of phenomena which I mentioned to you under the name of Hybridism, and which I explained to consist in the sterility of the offspring of certain species when crossed one with another. It matters not one whit whether this sterility is universal, or whether it exists only in a single case. Every hypothesis is bound to explain, or, at any rate, not be inconsistent with, the whole of the facts which it professes to account for; and if there is a single one of these facts which can be shown to be inconsistent with (I do not merely mean inexplicable by, but contrary to,) the hypothesis, the hypothesis falls to the ground,—it is worth nothing. One fact with which it is positively inconsistent is worth as much, and as powerful in negativing the hypothesis, as five hundred. If I am right in thus defining the obligations of an hypothesis, Mr. Darwin, in order to place his views beyond the reach of all possible assault, ought to be able to demonstrate the possibility of developing from a particular stock, by selective breeding, two forms, which should either be unable to cross one with another, or whose cross-bred offspring should be infertile with one another.

For, you see, if you have not done that you have not strictly fulfilled all the conditions of the problem; you have not shown that you can produce, by the cause assumed, all the phenomena which you have in nature. Here are the phenomena of Hybridism staring you in the face, and you cannot say, "I can, by selective modification, produce these same results." Now, it is admitted on all hands that, at present, so far as experiments have gone, it has not been found possible to produce this complete physiological divergence by selective breeding. I stated this very clearly before, and I now refer to the point, because, if it could be proved, not only that this *has* not been done, but that it *cannot* be done; if it could be demonstrated that it is impossible to breed selectively, from any stock, a form which shall not breed with another, produced from the same stock; and if we were shown that this must be the necessary and inevitable result of all experiments, I hold that Mr. Darwin's hypothesis would be utterly shattered.

But has this been done? or what is really the state of the case? It is simply that, so far as we have gone yet with our breeding, we have not produced from a common stock two breeds which are not more or less fertile with one another.

I do not know that there is a single fact which would justify any one in saying that any degree of sterility has been observed between breeds absolutely known to have been produced by selective breeding from a common stock. On the other hand, I do not know that there is a single fact which can justify any one in asserting that such sterility cannot be produced by proper experimentation. For my own part, I see every reason to believe that it may, and will be so produced. For, as Mr. Darwin has very properly urged, when we consider the phenomena of sterility, we find they are most capricious; we do not know what it is that the sterility depends on. There are some animals which will not breed in captivity; whether it arises from the simple fact of their being shut up and deprived of their liberty, or not, we do not know, but they certainly will not breed. What an astounding thing this is, to find one of the most important of all functions annihilated by mere imprisonment!

So, again, there are cases known of animals which have been thought by naturalists to be undoubted species, which have yielded perfectly fertile hybrids; while there are other species which present what everybody believes to be varieties * which are more or less infertile with one another. There are other cases which are truly extraordinary; there is one, for example, which has been carefully examined,—of two kinds of sea-weed, of which the male element of the one, which we may call A, fertilizes the female element of the other, B; while the male element of B will not fertilize the female element of A; so that, while the former experiment seems to show us that they are *varieties*, the latter leads to the conviction that they are *species*.

When we see how capricious and uncertain this sterility is, how unknown the conditions on which it depends, I say that we have no right to affirm that those conditions will not be better understood by and by, and we have no ground for supposing that we may not be able to experiment so as to obtain that crucial result which I mentioned just now. So that though Mr. Darwin's hypothesis does not completely extricate us from this difficulty at present, we have not the least right to say it will not do so.

There is a wide gulf between the thing you cannot

---

* And as I conceive with very good reason; but if any objector urges that we cannot prove that they have been produced by artificial or natural selection, the objection must be admitted—ultra-sceptical as it is. But in science, scepticism is a duty.

explain and the thing that upsets you altogether. There is hardly any hypothesis in this world which has not some fact in connection with it which has not been explained, but that is a very different affair to a fact that entirely opposes your hypothesis; in this case all you can say is, that your hypothesis is in the same position as a good many others.

Now, as to the third test, that there are no other causes competent to explain the phenomena, I explained to you that one should be able to say of an hypothesis, that no other known causes than those supposed by it are competent to give rise to the phenomena. Here, I think, Mr. Darwin's view is pretty strong. I really believe that the alternative is either Darwinism or nothing, for I do not know of any rational conception or theory of the organic universe which has any scientific position at all beside Mr. Darwin's. I do not know of any proposition that has been put before us with the intention of explaining the phenomena of organic nature, which has in its favour a thousandth part of the evidence which may be adduced in favour of Mr. Darwin's views. Whatever may be the objections to his views, certainly all other theories are absolutely out of court.

Take the Lamarckian hypothesis, for example. Lamarck was a great naturalist, and to a certain extent went the right way to work; he argued from what was undoubtedly a true cause of some of the phenomena of organic nature. He said it is a matter of experience that an animal may be modified more or less in consequence of its desires and consequent actions. Thus, if a man exercise himself as a blacksmith, his arms will become strong and muscular; such organic modification is a result of this particular action and exercise. Lamarck thought that by a very simple supposition based on this truth he could explain the origin of the various animal species : he said, for example, that the short-legged birds which live on fish, had been converted into the long-legged waders by desiring to get the fish without wetting their feathers, and so stretching their legs more and more through successive generations. If Lamarck could have shown experimentally, that even races of animals could be produced in this way, there might have been some ground for his speculations. But he could show nothing of the kind, and his hypothesis has pretty well dropped into oblivion, as it deserved to do. I said in an earlier lecture that there are hypotheses and hypotheses, and when people tell you that Mr. Darwin's strongly-based hypothesis is nothing but a mere modification of Lamarck's, you will know what to think of their capacity for forming a judgment on this subject.

But you must recollect that when I say I think it is either Mr. Darwin's hypothesis or nothing; that either we must take his view, or look upon the whole of organic nature as an enigma, the meaning of which is wholly hidden from us; you must understand that I mean that I accept it provisionally, in exactly the same way as I accept any other hypothesis. Men of science do not pledge themselves to creeds; they are bound by articles of no sort; there is not a single belief that it is not a bounden duty with them to hold with a light hand and to part with it, cheerfully, the moment it is really proved to be contrary to any fact, great or small.

And if in course of time I see good reasons for such a proceeding, I shall have no hesitation in coming before you, and pointing out any change in my opinion without finding the slightest occasion to blush for so doing. So I say that we accept this view as we accept any other, so long as it will help us, and we feel bound to retain it only so long as it will serve our great purpose—the improvement of Man's estate and the widening of his knowledge. The moment this, or any other conception, ceases to be useful for these purposes, away with it to the four winds; we care not what becomes of it!

But to say truth, although it has been my business to attend closely to the controversies roused by the publication of Mr. Darwin's book, I think that not one of the enormous mass of objections and obstacles which have been raised is of any very great value, except that sterility case which I brought before you just now. All the rest are misunderstandings of some sort, arising either from prejudice, or want of knowledge, or still more from want of patience and care in reading the work.

For you must recollect that it is not a book to be read, with as much ease, as its pleasant style may lead you to imagine. You spin through it as if it were a novel the first time you read it, and think you know all about it; the second time you read it you think you know rather less about it; and the third time, you are amazed to find how little you have really apprehended its vast scope and objects. I can positively say that I never take it up without finding in it some new view, or light, or suggestion that I have not noticed before. That is the best characteristic of a thorough and profound book;

and I believe this feature of the "Origin of Species" explains why so many persons have ventured to pass judgment and criticisms upon it which are by no means worth the paper they are written on.

Before concluding these lectures there is one point to which I must advert,—though, as Mr. Darwin has said nothing about man in his book, it concerns myself rather than him;—for I have strongly maintained on sundry occasions that if Mr. Darwin's views are sound, they apply as much to man as to the lower mammals, seeing that it is perfectly demonstrable that the structural differences which separate man from the apes are not greater than those which separate some apes from others. There cannot be the slightest doubt in the world that the argument which applies to the improvement of the horse from an earlier stock, or of ape from ape, applies to the improvement of man from some simpler and lower stock than man. There is not a single faculty—functional or structural, moral, intellectual, or instinctive,—there is no faculty whatever that is not capable of improvement; there is no faculty whatsoever which does not depend upon structure, and as structure tends to vary, it is capable of being improved.

Well, I have taken a good deal of pains at various times to prove this, and I have endeavoured to meet the objections of those who maintain, that the structural differences between man and the lower animals are of so vast a character and enormous extent, that even if Mr. Darwin's views are correct, you cannot imagine this particular modification to take place. It is, in fact, easy matter to prove that, so far as structure is concerned, man differs to no greater extent from the animals which are immediately below him than these do from other members of the same order. Upon the other hand, there is no one who estimates more highly than I do the dignity of human nature, and the width of the gulf in intellectual and moral matters, which lies between man and the whole of the lower creation.

But I find this very argument brought forward vehemently by some. "You say that man has proceeded from a modification of some lower animal, and you take pains to prove that the structural differences which are said to exist in his brain do not exist at all, and you teach that all functions, intellectual, moral, and others, are the expression or the result, in the long run, of structures, and of the molecular forces which they exert." It is quite true that I do so.

"Well, but," I am told at once, somewhat triumphantly, "you say in the same breath that there is a great moral and intellectual chasm between man and the lower animals. How is this possible when you declare that moral and intellectual characteristics depend on structure, and yet tell us that there is no such gulf between the structure of man and that of the lower animals?"

I think that objection is based upon a misconception of the real relations which exist between structure and function, between mechanism and work. Function is the expression of molecular forces and arrangements no doubt; but, does it follow from this, that variation in function so depends upon variation in structure that the former is always exactly proportioned to the latter? If there is no such relation, if the variation in function which follows on a variation in structure, may be enormously greater than the variation of the structure, then, you see, the objection falls to the ground.

Take a couple of watches—made by the same maker, and as completely alike as possible; set them upon the table, and the function of each—which is its rate of going—will be performed in the same manner, and you shall be able to distinguish no difference between them; but let me take a pair of pincers, and if my hand is steady enough to do it, let me just lightly crush together the bearings of the balance-wheel, or force to a slightly different angle the teeth of the escapement of one of them, and of course you know the immediate result will be that the watch, so treated, from that moment will cease to go. But what proportion is there between the structural alteration and the functional result? Is it not perfectly obvious that the alteration is of the minutest kind, yet that slight as it is, it has produced an infinite difference in the performance of the functions of these two instruments?

Well, now, apply that to the present question. What is it that constitutes and makes man what he is? What is it but his power of language—that language giving him the means of recording his experience—making every generation somewhat wiser than its predecessor,—more in accordance with the established order of the universe?

What is it but this power of speech, of recording experience, which enables men to be men—looking before and after and, in some dim sense, understanding the working of this wondrous universe—and which distinguishes man from the whole of the brute world?

I say that this functional difference is vast, unfathomable, and truly infinite in its consequences; and I say at the same time, that it may depend upon structural differences which shall be absolutely inappreciable to us with our present means of investigation. What is this very speech that we are talking about? I am speaking to you at this moment, but if you were to alter, in the minutest degree, the proportion of the nervous forces now active in the two nerves which supply the muscles of my glottis, I should become suddenly dumb. The voice is produced only so long as the vocal chords are parallel; and these are parallel only so long as certain muscles contract with exact equality; and that again depends on the equality of action of those two nerves I spoke of. So that a change of the minutest kind in the structure of one of these nerves, or in the structure of the part in which it originates, or of the supply of blood to that part, or of one of the muscles to which it is distributed, might render all of us dumb. But a race of dumb men, deprived of all communication with those who could speak, would be little indeed removed from the brutes. And the moral and intellectual difference between them and ourselves would be practically infinite, though the naturalist should not be able to find a single shadow of even specific structural difference.

But let me dismiss this question now, and, in conclusion, let me say that you may go away with it as my mature conviction, that Mr. Darwin's work is the greatest contribution which has been made to biological science since the publication of the "Règne Animal" of Cuvier, and since that of the "History of

Development," of Von Baer. I believe that if you strip it of its theoretical part it still remains one of the greatest encyclopædias of biological doctrine that any one man ever brought forth; and I believe that, if you take it as the embodiment of an hypothesis, it is destined to be the guide of biological and psychological speculation for the next three or four generations.

THE END.

ROBERT HARDWICKE, PRINTER, 192, PICCADILLY.

www.ingramcontent.com/pod-product-compliance
Lightning Source LLC
Chambersburg PA
CBHW030310170426
43202CB00009B/943